J Herzfeld

The Technical Testing of Yarns and Textile Fabrics

With Reference to Official Specifications

J Herzfeld

The Technical Testing of Yarns and Textile Fabrics
With Reference to Official Specifications

ISBN/EAN: 9783743687066

Printed in Europe, USA, Canada, Australia, Japan

Cover: Foto ©berggeist007 / pixelio.de

More available books at **www.hansebooks.com**

THE TECHNICAL TESTING

OF

YARNS AND TEXTILE FABRICS

WITH REFERENCE TO OFFICIAL SPECIFICATIONS

BY

Dr. J. HERZFELD

PRINCIPAL OF THE CHEM.-TECHN. EXPERIMENTAL AND TEACHING INSTITUTE,
COLOGNE; FORMERLY INSTRUCTOR AT THE ROYAL HIGH SCHOOL
FOR WEAVING, MÜLHEIM, ETC.

WITH SIXTY-NINE ILLUSTRATIONS

TRANSLATED BY

CHAS. SALTER

LONDON
SCOTT, GREENWOOD & CO.
PUBLISHERS OF THE
Hatter's Gazette, Oil Colourman's Journal, etc.
19, 21 AND 23 LUDGATE HILL, CITY, E.C.
1898

PREFACE.

CONSIDERABLE attention has always been devoted to the technical testing of yarns and textile fabrics, as the numerous processes employed in practice testify. It was always incumbent on the spinner to control the progress of the spinning process by occasional tests; and the weaver on his part subjected to examination—though very often by the hand and eye alone—the material to be worked up. In the last decade new combinations have been effected in yarns, new textile fibres come to the front, and certain deceptions designedly practised, all of which circumstances have so increased the importance, to the spinner and weaver, of submitting the articles to examination, that a knowledge of the methods has become more than ever indispensable. The requirements exacted, now-a-days, with regard to the various qualities of the goods render it essential for the manufacturer to minutely test both the raw materials, the intermediate products and the finished article; since not only the public but also (and principally) the large purchasers of textile fabrics, and the various administrative bodies, such as Army Clothing Departments, Railway Companies, etc., etc., have adopted definite specifications to ensure having good material and workmanship, compliance with which is a necessary condition for the acceptance of the articles supplied. The Customs officials have their processes on which they base their decision as to the duties payable, and the wholesale merchant and retail dealer insist on various conditions when purchasing, in order to comply with the increasing exactions of the consumer.

In the present work are collected all the tests, both of a physical and chemical nature, that can, for technical purposes, be performed on yarn or cloth, so that not only the commercial and textile chemist who has frequently to reply to questions of this kind, but also (and principally) the practical manufacturer of textiles and his subordinates, officials, overlookers, the spinner, the weaver, the dyer and the finisher are catered for.

Moreover, all those who have to do with the buying and selling of textile goods will be enabled to learn and apply to considerable advantage the various processes described.

The appendix, containing the latest official testing specifications for Army clothing and supplies, will be useful to those manufacturers and merchants who give their attention to this class of business.

TRANSLATOR'S PREFACE.

In the present work the author has gathered together, probably for the first time, a full collection of the various tests that may, at one time or another, have to be imposed on yarns or cloths, and has illustrated those in which mechanical force is employed by typical machines.

Moreover, he has given us a comparative scheme for the conversion of English yarn numbers into their corresponding values, as well as information regarding the various continental methods of reeling, both of which should prove useful to the English manufacturer doing an export trade, and help to smooth some of the complications arising in the conversion of our own complex system of weights and measures into the metric system, a task of peculiar difficulty—as the writer is, from personal experience, aware—in the textile business.

These considerations will, it is hoped, render the work serviceable to a large circle of readers in this very important branch of industry, as well as to the textile chemist, whose business it is to detect the presence of extraneous fibres in yarn and cloth, estimate the loading of silk, and determine the colouring matters that may have been used in dyeing the samples submitted to him for examination.

TABLE OF CONTENTS.

	PAGE
Preface	v
YARN TESTING	1
I. Microscopic Examination of Textile Fibres	3
The microscope: its parts, manipulation and testing	3
1. Vegetable fibres	14
(a) Cotton	16
(b) Flax, Linen	18
(c) Hemp	20
(d) Nettle fibre, China grass (Ramie)	22
(e) Jute	23
Manila hemp, New Zealand flax, cocoanut fibre, Cosmos fibre	25-26
2. Wool	27
(a) Sheep's wool	29
(b) Mohair wool	30
(c) Alpaca wool	31
Cashmere, Vicuña, llama, camel hair, hare and rabbit fur, horse hair, cows and calves' hair	31-33
3. Artificial wools (shoddy, mungo, extract)	33
Microscopical examination	34
Chemical examination	38
4. Silk	39
Silk from Bombyx mori	41
Chappe silk	41
Tussah silk	41
Loaded silk	42
Artificial silk	43
II. Chemical Examination of Textile Fibres	43
Preparation of test solutions	44
1. Characteristic coloration by dye stuffs	48
2. Influence of various salts in solution	48
3. ,, ,, alkaline liquids, etc.	49

	PAGE
4. Influence of acids, metalloids, etc.	50
Analytical table for examining mixed fibres	51
Qualitative and quantitative estimations	52
Quantitative estimation of the loading of silk	55
III. Determining the Yarn Number	57
Variations from standard	58
1. Cotton	58
2. Linen yarns	62
3. Jute yarns	64
4. Ramie, nettle fibre	65
5. Wool	66
6. Silk	69
7. Chappe silk	71
Apparatus for ascertaining the yarn number	72
Sampling reels	77
IV. Testing the Length of Yarns	79
V. Examination of the External Appearance of Yarn	81
Yarn tester	81
VI. Determining the Twist of Yarn and Twist	83
(a) Cotton yarns	84
(b) Linen yarns	86
(c) Woollen yarns	86
(d) Silk	88
Apparatus for determining the twist of yarn	90
VII. Determination of Tensile Strength and Elasticity	93
Apparatus for testing the breaking strain of yarn	97
VIII. Estimating the Percentage of Fat in Yarn	107
IX. Determination of Moisture (conditioning)	108
Conditioning silk	110
,, loose wool	111
,, worsted	113
Apparatus	113
TESTING MANUFACTURED FABRICS	118
1. Plain or smooth weavings	118
2. Twilled fabrics	119
3. Figured fabrics	119
4. Velvety fabrics	120
Classification of woven goods	120
The testing of fabrics	123

TABLE OF CONTENTS.

	PAGE
I. Determination of the Mode of Weaving, Distinction and Combination of Warp and Weft Threads	124
II. Testing the Strength and Elasticity of a Fabric	125
Breaking strain testers	128
(a) Rehse's tester	128
(b) Leuner's „	128
(c) Breaking strain and elasticity tester	130
(d) Tarnagrocki's tester	132
III. Ascertaining the Count of Warp and Weft Threads in a Fabric	141
IV. Determination of Shrinkage	143
Determining the thickness of cloth	143
V. Examining the Constituents of the Warp and Weft. Weighing	144
Weighing on the balance	144
Determining the weight of cloth from the count and the yarn number	144
Determining the weight of the individual constituents of the cloth. Quantitative chemical analysis of the fabric	145
1. Mixed fabrics containing wool and cotton	145
2. „ „ of cotton and silk	146
3. „ „ of wool and silk	147
4. „ „ of cotton, wool and silk	147
VI. Determination of the Dressing	148
1. Physical examination	150
2. Chemical „	150
VII. Estimation of the Waterproof Properties of Cloth	154
VIII. Determining Hygroscopicity	157
IX. Testing the Fastness of the Dye	158
(a) Washing fastness	159
(b) Fastness under friction	160
(c) Resistance to perspiration	160
(d) Fastness against rain	161
(e) Resistance to street mud and dust	161
(f) Fastness to weather, light and air	162
(g) Resistance to ironing and steaming	163
X. Measuring the Length of Piece Goods	164
XI. Determination of Mordants and Dyes	165

	PAGE
1. Blue and violet dyes	169
Testing for pure indigo dyes	169
2. Green dyes	170
3. Red and red-brown dyes	171
4. Yellow and orange dyes	172
5. Browns, greys and mode shades	172
6. Black dyes	178
XII. Detection and Estimation of Arsenic	178

APPENDIX.

	PAGE
Official specifications for the supply of materials for use in the German Army	181
1. Cloth	181
2. Linens and cottons for military use	187
(a) Linens	187
(b) Cottons	188
3. Waterproof materials for military purposes	192
(a) Canvas for laced shoes	192
(b) Stuffs for provision bags and knapsacks	196
(c) Tent canvas (for portable tents)	199
4. Linens and cottons for barrack and hospital use and for the training colleges	200
A. Linens	201
B. Cottons	202
C. Woollens (flannel)	202
Contract specifications for French Army clothing	203
INDEX	205

TABLE OF ILLUSTRATIONS.

		PAGE
Fig. 1.	Microscope for laboratory purposes	4
„ 2.	„ with micrometer screw fine adjustment	4
„ 3.	Revolving carrier for two objectives	6
„ 4.	Hyparchia janira. Scale of butterfly wing	8
„ 5.	Pleurosigma angulatum. Test object	9
„ 6.	Camera lucida	10
„ 7.	Abbe's drawing apparatus	11
„ 8.	Ocular micrometer	11
„ 9.	Polariser	12
„ 10.	Magnified cotton fibres	17
„ 11.	Sectional view of cotton fibres	17
„ 12.	Section of "dead" or immature cotton	17
„ 13.	Magnified flax fibres	19
„ 14.	Cross section of flax fibres	19
„ 15.	Magnified fibres of hemp	21
„ 16.	Cross section of hemp fibres	21
„ 17.	Magnified nettle fibres	22
„ 18.	Cross section of nettle fibres	23
„ 19.	Magnified jute fibres	24
„ 20.	Cross section of jute fibres	24
„ 21.	Magnified fibres of wool	29
„ 22.	Cross section of wool fibres	30
„ 23.	Camel wool	32
„ 24.	Shoddies	34
„ 25.	Shoddy yarns (magn. seventy-fold)	36
„ 26.	Silk fibres	40
„ 27.	Organzine silk	40
„ 28.	Tussah silk	42
„ 29.	Loaded silk	42
„ 30.	Arc balance	72
„ 31.	Micrometric balance	73
„ 32.	Precision balance	74

TABLE OF ILLUSTRATIONS.

		PAGE
Fig. 33.	Dietze's yarn balance	75
„ 34.	Steel balance	76
„ 35.	Sampling reel	77
„ 36.	„ „	78
„ 37.	Counting reel	80
„ 38.	Yarn tester	82
„ 39.	„ „	82
„ 40.	Twist tester	90
„ 41.	„ „	90
„ 42.	Heal's twist tester	91
„ 43.	Twist tester with expansion measurer and turn counter	92
„ 44.	Pocket instrument for testing strength of yarn	98
„ 45.	„ „ „ „ „	98
„ 46.	Breaking strain and elasticity tester	98
„ 47 and 48.	Breaking strain tester	99
„ 49.	Elasticity tester	100
„ 50.	Combined breaking strain and elasticity tester	100
„ 51.	Breaking strain tester	100
„ 52.	Piat and Pierrel's breaking strain tester	101
„ 53.	Schopper's breaking strain tester	103
„ 54.	Breaking strain and elasticity tester	104
„ 55.	Clamps	104
„ 56.	Holzach's continuous tester	106
„ 57.	Conditioning apparatus	110
„ 58.	„ „ sectional view	110
„ 59.	Kohl's wool conditioning apparatus	113
„ 60.	Heal's conditioning oven	114
„ 61.	Ulmann's conditioning apparatus	115
„ 62.	Conditioning apparatus with sliding weight	116
„ 63.	Leuner's cloth tester	129
„ 64.	Breaking strain and elasticity tester	130
„ 65.	Tarnagrocki's cloth tester	131
„ 66.	„ „ „	133
„ 67.	Horizontal cloth tester	134
„ 68.	Gawalowski's waterproof tester	155
„ 69.	Cloth measuring machine	164

THE TECHNICAL TESTING

OF

YARNS AND TEXTILE FABRICS.

YARN TESTING.

By the term "yarn" is understood the final product of the process of spinning, or the union of textile fibres into a single thread. According to the nature of the raw material employed we may have cotton yarn, linen yarn, jute yarn, carded (wool) yarn, worsted yarn, etc., etc. Union (half wool) yarn is a mixture of wool and cotton (containing 15 per cent., or sometimes, particularly for export, as much as 50 to 60 per cent., of cotton). Vicuna yarn is a mixture of 90 per cent. and upwards of cotton with 3 to 10 per cent. of wool, and "imitation" yarn consists solely of cotton. By silk yarns, only those prepared by spinning waste silk (floret) are understood. An artificial woollen yarn may either be composed of shoddy alone or of admixtures of this material with good wool or even cotton.

Yarns are classified, according to the method of their employment in weaving, as warp or weft; according to the mode of preparation, as bleached or unbleached, dyed or undyed, printed, dressed and loaded yarn; and by the way they are made up for use, in cops, conets, bobbins, hanks, or balls, etc.

The union of several threads constitutes a "twist," which,

apart from its use as a warp in weaving, finds employment for sewing, knitting and embroidery purposes. According to the number of the threads taken, we have two-thread, three-thread twist, and so on, but it is seldom that the number exceeds eight. For cords, ropes and cables, however, several strands of twist are frequently joined together by twisting.

Distinction must be drawn between the "twisted" yarn produced by tightly twisting together the individual threads, and the "doubled" yarns, consisting of slightly twisted twist.

A spun thread is perfect when its diameter is regular throughout, *i.e.*, free from knots or weak places, with the proper degree of twist and resistance to breakage, the surface being rough or smooth according to the material and its intended use.

The examination and testing of a yarn must therefore extend to the following considerations :—

1. The microscopical and chemical examination of the raw material. According to the result of this the quantitative estimation of the individual threads of mixed yarns is proceeded with; and in the case of silk, the amount of loading.

2. Determination of the "number" of the yarn. In many cases a comparison of various yarns ensures sufficient accuracy.

3. Measuring the length of yarn in the hanks, cops, bobbins, balls, etc.

4. Testing the twist of the thread. The degree of twist is a definite quantity dependent on the fineness of the yarn, the length of the fibres, and the use of the yarn (*i.e.*, as warp or weft). In silk, the original twist of the cocoon fibre and the secondary twist are determined.

5. Examination of the external appearance of the yarn; whether a sleek, smooth worsted yarn, or a more woolly carded yarn; a wet or dry-spun flax yarn; whether evenly

spun (not "tapering"); and whether the threads have been singed, or smoothed, or sized by machinery, etc.

6. Testing the breaking strain and stretch of a yarn. These properties, depending on the strength and length of the fibres, as well as on the degree of twist, were in the past frequently tested merely by hand, but the end is more satisfactorily attained by the use of the dynamometer.

7. A knowledge of the percentage of fat is, as a rule, desired only in the carded and "artificial" wool yarns.

8. Estimation of the degree of moisture (hygroscopicity). This test, up to now performed almost exclusively on silk, is latterly coming into use for wool and woollen yarns as well.

I. MICROSCOPIC EXAMINATION OF TEXTILE FIBRES.

THE MICROSCOPE: ITS PARTS, MANIPULATION AND TESTING.

For examining the yarn with a view to detecting the raw materials from which it has been spun, the so-called compound microscope is employed. This consists principally of a tube closed at the upper end by a large, and at the lower end by a smaller, glass lens, with different focal lengths.

An object may be examined under a lens in two ways: either by bringing it within or beyond the focal length of the lens. In the former case, as is seen in the simple magnifying glass, an enlarged picture is obtained on the side next the object; but when the latter is at a distance greater than the focal length, the enlarged picture is formed in an inverted position on the opposite side of the lens. In the compound microscope both these conditions are combined, for the purpose of obtaining a very high power of magnification.

Of the two lenses referred to as terminating the microscope tube, the larger one with the greater focal length is placed next the eye and the smaller one of lesser focal length nearest the object. The latter is therefore termed the objective and the former the eye-piece.

Eye-piece and objective are fitted in a tube some six to

 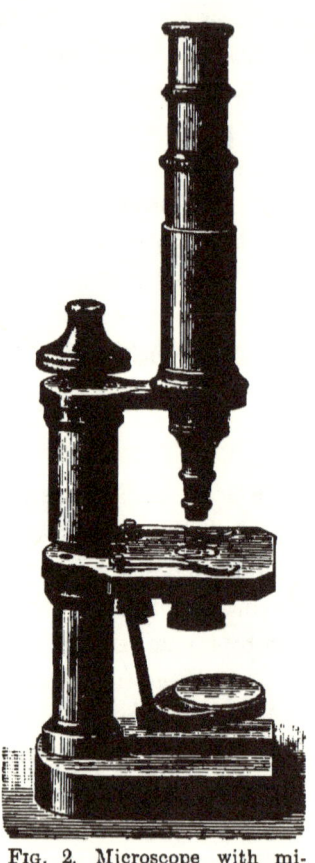

Fig. 1. Microscope for laboratory and technical purposes, with rack and pinion adjustment.

Fig. 2. Microscope with micrometer screw fine adjustment. Cylindrical substage diaphragm.

seven inches in length, capable of vertical movement, and blackened on the inside to exclude extraneous light. The object is first observed at a distance greater than the focal length of the objective, which for this reason is kept very small, whereupon an inverted magnified picture is projected at a certain distance inside the tube. At this place an annular diaphragm is inserted in the tube in order to limit the field of vision and exclude the circumferential rays tending to diminish the clearness of the picture. The length of the tube is so calculated that this picture falls within the focal length of the eye-piece at the upper end, whereby a re-enlargement of the picture is effected. The total magnifying power of a microscope is therefore the sum of the powers of the objective and the eye-piece.

To ensure clearness (definition) of the picture it is advisable to produce magnification chiefly by means of the objective. The latter is, as a rule, composed of several achromatic double lenses, the eye-piece consisting of a system composed of the true ocular lens and a collecting lens, the object of the latter being to enlarge the field of vision and increase the definition of the picture, even though the size of the latter be simultaneously reduced. Were this lens not employed a portion of the rays proceeding from the picture would escape the eye-piece.

The tube carrying the eye-piece and objective can be raised and lowered by the hand (though for beginners the rack and pinion motion is more certain) to regulate the coarse or approximate adjustment, whilst the fine adjustment is effected by the so-called micrometer screw. On the stand is fixed an arrangement for supporting the stage, as well as one for supplying the necessary light to the object under examination, *viz.*, the substage condenser, a small circular concave reflector, movable in any direction. The stage is pierced with a small circular aperture for the passage of the reflected light, and may be either fixed or movable. The concentration of the light rays and the cutting off of the circumferential rays is effected by a revolving metal diaphragm fitted below the stage and perforated by several apertures of different diameters, their centres being equidistant from that of the diaphragm. For higher powers a cylindrical diaphragm is employed, consisting of a hollow cylinder, blackened on the inside, and affixed to the aperture in the stage. The best source of illumination is diffuse daylight with a sky evenly covered with a white veil of clouds. For evening work it is advisable to employ a glass bulb filled with a dark blue solution of ammoniacal copper oxide interposed between the source of light and the condenser.[1] A solid foundation is afforded the microscope by making the base of the stand in the shape of a horse shoe.

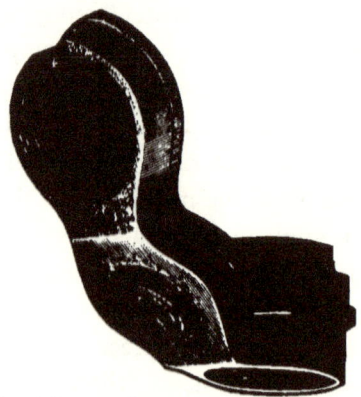

Fig. 3. Revolving carrier (nose-piece) for two objectives to screw on to the body.

When an object is to be examined by objectives of various powers in succession, use is made of a revolving objective-carrier or nose-piece screwed on to the lower end of the body tube, this arrangement facilitating the rapid change of powers.

The picture produced by the microscope must be very clear at the edges; the structural relationship of the details of the object should be well defined (analytical power), and the magnifying power must not be too low. This latter

[1] *Translator's Note.*—The various incandescent gas lights give a very good white light for microscope work.

faculty can be easily measured by observing a very fine scale (*e.g.*, a millimetre graduated to 100 divisions) on a glass slide under the microscope. This task will, however, have already been performed by the maker of the instrument, and will be denoted in a suitable manner by means of letters and figures on the objective and eye-piece.

The eye-pieces are numbered 1 to 5 :—

Number of eye-piece - - -	1	2	3	4	5
Focal length in *mm*. - - -	50	40	30	25	20
Magnifying power - - - -	3	4	5·5	7	9

The objectives are marked with letters :—
a_1, a_2, a_3, aa, A, AA, B, C, D, DD, E, F, etc.

The medium objectives of a focal length between 18 and 4·3 *mm*. are constructed with larger (indicated by double letters) or smaller apertures, the former indicating greater analytical power.

The linear magnifying powers of the objectives in combination with the eye-pieces are given (for a 160 *mm*.—6¼ in.—tube) in the following table :—

Eye-piece.	Objective.					Eye-piece.	Focal Length in *mm*.
	1	2	3	4	5		
a_1	7	10	15	20	—	a_1	40
a_2	11	16	23	30	—	a_2	35
a_3	20	30	40	50	—	a_3	30
aa	25	35	47	60	77	aa	26
A, AA	37	50	70	90	115	A, AA	18
B	60	85	115	145	185	B	12
C	105	145	200	265	325	C	7
D, DD	175	240	325	420	540	D, DD	4·3
E	280	390	535	680	865	E	2·7
F	415	585	790	1000	1275	F	1·85

8   YARNS AND TEXTILE FABRICS.

The power of each combination can thus be seen at a glance.

In judging the quality of a microscope the magnifying power is not the sole consideration, the most important being the capacity of the lenses to produce a sharply defined and clear picture plainly showing all details of structure. The testing of the instrument is performed by examining sundry test objects, such as small parts of animal or vegetable organisms prepared in a suitable manner. For low powers grains of potato starch suspended in water or glycerine form a suitable test object. The individual layers of the grain arranged around the eccentric hilum must be clearly revealed in sharp, bold and delicate outline. For

FIG. 4. Hyparchia janira.   Scale of butterfly wing.   Test object for all systems.

testing higher, and the very highest powers, the wing scales of the butterfly (Hyparchia janira) and the siliceous plates of an alga (Pleurosigma angulatum) are most frequently employed.

(a) *Hyparchia janira.*—The scale is rectangular in shape, with three broad points at the upper end, the surface being covered with 22 to 24 longitudinal striations.

Even with a magnifying power of 300 times, a large number of very delicate cross stripes can be observed between the striations.

(b) *Pleurosigma angulatum.*—The siliceous skeleton of this alga exhibits a central ridge. The first indications of mark-

MICROSCOPIC EXAMINATION OF TEXTILE FIBRES. 9

ings become apparent with a magnifying power of 200 to 250 times, and with a 300-power the markings are well defined and three systems of striation are recognisable. When the power is increased a number of laterally arranged hexagons are revealed.

Supplementary Apparatus to the Microscope.—For microscope work various additional appliances are frequently required.

(*a*) *The Dissecting Microscope.*—This arrangement consists

FIG. 5. Pleurosigma angulatum. Test object for the analytical power of medium and high powers : 200-400 ; 400-800 times.

of a stand on which is mounted a simple lens. At a distance equal to the focal lens of the glass is situated a stage for supporting the object to be dissected, which, with the assistance afforded by the low power lens, can be treated with small needles and knives, and prepared for examination under the large microscope. The necessary light is reflected by a small condenser below the stage. When the operator's eyesight is good this apparatus may be dispensed with. In dissecting yarn a sheet of black glazed paper is spread on the work-table, the tiny fibres being more plainly visible and more readily seized on such a background.

(b) *The Camera Lucida.*—Various appliances have been constructed for copying microscopic pictures, by projecting the image, by means of glass prisms (Fig. 6) or reflectors (Fig. 7), on to an adjacent sheet of paper, so that the outlines can be reproduced by pencilling. It is necessary to have the paper on a level with the microscope stage. Apparatus with this object have been designed by Wollaston,

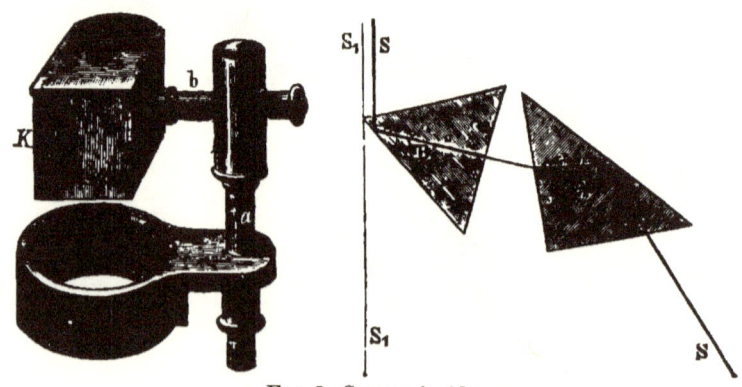

Fig. 6. Camera lucida.

Nobert, Sömering and others, those in the appended illustrations being, respectively, the camera lucida and the Abbe apparatus. To obtain pictures free from distortion it is necessary that the drawing surface should be inclined at an angle of 25°.

(c) *The Micrometer.*—Various methods are adopted for measuring the diameters of fibres, the simplest being the glass micrometer, a fine scale engraved on glass. The measurement is performed either on the object itself (objective micrometer) or on the image (ocular micrometer). The former consists of a glass slide on which is engraved a millimetre graduated into 100 equal parts, each of which is termed a micromillimetre (1 $mmm.$, or $\mu = 0.001$ $mm.$). A useful form is the ocular micrometer, which consists of either an arbitrary scale or of 1 centimetre divided into 100 parts (or $\frac{1}{2}$ $cm.$ with 50 divisions). In the former

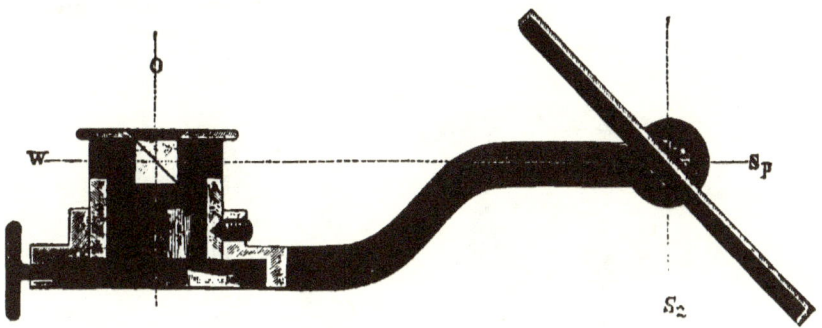

FIG. 7. Abbe's drawing apparatus.

event, the value of the scale must be ascertained by comparison with an objective micrometer, which is laid on the stage, whilst the ocular scale is inserted in the diaphragm of the eye-piece to determine how many of the divisions of the latter scale coincide with those on the former.

The Polarising Apparatus.—This apparatus is advisable for

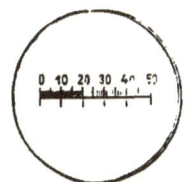

FIG. 8. Ocular micrometer.

use in the microscopic examination and differentiation of the various kinds of silk, and probably for vegetable fibres as well. It consists of two parts: a prism (the polariser), which is placed below the object on the stage of the microscope and allows only straight rays of polarised light to pass from the reflector through the object; and a second prism (the analyser) placed above the object—preferably on top of the eyepiece—and serving to analyse the rays of light transmitted through it. In employing this apparatus only very low powers are used.

Fig. 9. Polariser. To be suspended in the illuminating apparatus.

Making the Preparations.—Yarns to be subjected to microscopic examination are first of all dissected into fine fibres after removal by suitable means (referred to later) of all dirt, colouring matter, etc., so that the passage of the light will be unrestricted. It is advisable to immerse the object in water, or, better still, in a liquid like glycerine or Canada balsam, which will increase its transparency. A preliminary maceration of the fibres by prolonged boiling in water is very advantageous, and, in the case of vegetable fibres, boiling for a few seconds in nitric acid containing a little potassium chloride is recommended. The fibres are then laid on a glass slide (1 × 3 ins.), separated one from another, but arranged side by side, and covered with a small circular or square cover glass (15 to 24 *mm.* in diameter and 0·15 to 0·20 *mm.* in thickness). To permanently preserve the preparation a drop of glycerine jelly is laid on the slide before applying the cover glass and warmed slightly by means of a lighted match held a little way below the cover, whereupon the liquid distributes evenly, and after gently pressing down the cover glass the whole is set to cool. The edges of the cover glass are coated over with black varnish. The subsidiary instruments used in these operations consist of small knives (scalpels), needles, lancets, forceps and scissors.

Preparing Sections.—It is very often important to prepare sections in order to differentiate, for instance, between mature and immature cotton fibres, etc. To this end a number of the fibres are arranged parallel and embedded in melted tallow or paraffin, or a small bundle of the fibres can be impregnated with a thick solution of gum containing a little glycerine, the dried mass being firmly bound between two corks and a thin section cut by means of a smooth ground razor in a direction at right angles to the axis of the fibres. A special apparatus known as the microtome, in which the knife moves in a guide frame and cuts the preparation through at an acute angle, is also used.

The small sheet of paraffin is transferred to a slide, and, after being slightly warmed, the matrix is removed by turpentine or benzol. If the section is not sufficiently transparent it is steeped in glycerine, carbolic acid, etc.

The Microscopic Examination.—The simplest and best examination is performed by the microscope alone; in case of doubt a microchemical test is also introduced, by admitting one or two drops of certain liquids under the cover glass and observing under the microscope the reactions ensuing thereon.

In examining undyed fibres it is advisable, as already mentioned, to previously steep or boil them in water, or if, as in the case of wool, the fibres are contaminated by adherent fat, to remove the latter by boiling with alcohol or treatment with ether, carbon bisulphide, etc. Suitable reagents for increasing the transparency should also be used.

In the case of coloured fibres, the dressing and colouring matters should be removed by boiling in an alkaline or weak acid bath, or by extraction with alcohol, ether, etc.

1. VEGETABLE FIBRES.

The fibres employed in industrial processes differ considerably in point of anatomical structure. Those belonging to the group of plant hairs are almost exclusively seed hairs, or the hairy covering of the skin of the seed. Such are: cotton, bombax fibre and vegetable silk (Asclepias). Many fibres are composed of the vascular bundles of the leaves, stems, or roots of monocotyledonous plants, such as New Zealand flax, Manila hemp, etc. Most frequently, however, the bast fibres of dicotyledonous plants, such as hemp, flax, jute, China grass, etc., are utilised.

Plant hairs exhibit (apart from branchings) only a single apex of variable shape, and always appear to be coated externally with a thin skin or cuticle, which remains undissolved when treated with sulphuric acid. The cell walls may be thick or thin, structureless, porous, or reticulated. The chief content of the hollow interior, or lumen, consists of air. Frequently the hair is flattened in a band, so that the lumen is almost *nil*. The cross section is also highly characteristic.

Bast fibres are composed of enclosed tubes with pointed ends, mostly with stout walls and of rounded or elongated cross section. The inner wall is often covered with a very thin layer of strongly adherent dried protoplasm, but well-defined pores are rare. Highly characteristic are the so-called dislocations or tubercles, which, by their property of becoming more intensely coloured than the rest of the fibre when stained with iodochloride of zinc, can be readily detected. The terminals of the bast fibres are sharp pointed or blunt, simple or with branching points, thin or thick walled, etc.

The subjoined table affords a survey of the length of various fibres and the ratio of length to breadth in the cells:—

Fibre.	Length of Crude Fibres in cm.	Length of the Cells in cm.	Maximum Width of Cells.	
			Extreme Limits in μ.	Usual Width in μ.
Cotton :				
Sea Island (Gossypium barbadense)	4·05	4·05	19·2—27·9	25·2
Bengal (G. herbaceum)	1·82	1·82	11·9—22·0	18·9
Indian (G. arboreum)	2·50	2·50	20·0—37·8	29·9
Flax	20—140	2·0—4·0	12—25	16
Hemp	100—300	0·8—4·1	16—32	20
Jute	150—300	0·8—4·1	16—32	20
Nettle	—	8·0	16—126	—
New Zealand Flax	80—110	2·5—5·6	8—29	—

The chief chemical constituents of all fibres are cellulose and woody fibre, the former constituting the principal bulk, whereas the latter, which reduces the value of the fibre, is not contained in all. Its presence is determined by the following colour tests : Aniline salts, golden yellow ; phloroglucin and hydrochloric acid, red ; indol and hydrochloric acid, or phenol and hydrochloric acid, green ; zinc iodochloride, yellow to brown. Pure cellulose, such as cotton, stains blue with iodine and sulphuric acid, and violet with zinc iodochloride. The woody fibre is destroyed by bleaching, so that well bleached jute or hemp no longer gives the original colour reactions.

Finally, the phenomena of distension and polarisation are investigated by the aid of the microscope.

The vegetable fibres are distinguishable from those of animal origin by their behaviour in presence of acids and alkalis, the former being insoluble when boiled with soda or potash-lye, but readily soluble in sulphuric acid, and evolving no smell of horn when burnt.

(*a*) Cotton.

Cotton forms the seed hairs of various kinds of Gossypium, especially Gossypium herbaceum, G. arboreum, G. barbadense, etc. The fibre is unicellular, with variously formed apex, mostly thick walled. The diameter and length vary with the kind of cotton, the former ranging between 12 and 35 μ, and the mean length from 10 to 40 mm. Under the microscope the fibre appears as a wide finely-granulated band, frequently twisted round its own axis (particularly American cotton). The cell wall being very thin, the lumen consequently appears very large, generally about two-thirds of the total breadth. In spun fibres the twists are more elongated than in the crude state.

By "dead" cotton is understood such as has not attained full maturity. Its detection is very important, since its presence in yarn spoils the durability of the latter. It is recognisable by the very thin transparent threads, which, though band-shaped, are not twisted, and exhibit not the slightest trace of lumen. The cross section is most highly characteristic, looking like a collapsed tube with thick walls,

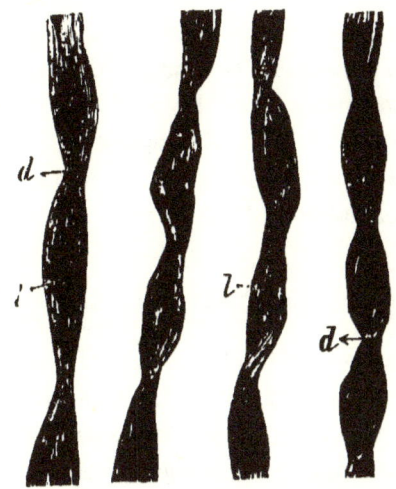

Fig. 10. Magnified cotton fibres: *l*, lumen; *d*, torsion.

Fig. 11. Sectional view of cotton fibres.

Fig. 12. Section of "dead" or immature cotton.

an appearance (Fig. 12) which it is hardly possible to confound with that of the mature fibre.

When cotton fibre is treated with ammoniacal copper oxide it exhibits a remarkable distension, and, finally, solution of the cellulose, only a ring of cuticle, or thin skin covering the cell wall, being left behind. This integument is thinner in the finer varieties of cotton than in other kinds. In well-bleached fibres distension may not occur, and the skin may be altogether lacking; in any case, a concentrated solution is necessary for the production of the first-named phenomenon.

Microchemical Reactions.—Iodine and sulphuric acid stain the fibre blue; madder tincture, red; fuchsine, red, this coloration disappearing on the addition of ammonia, and thus affording a means of distinguishing cotton from flax. Sulphuric acid rapidly effects solution, and concentrated soda lye causes a contraction of the internal space (mercerising).

(b) FLAX, LINEN.

Flax is composed of the bast fibres of the stem of the flax plant (Linum usitatissimum). Like cotton they consist of pure cellulose, are of regular thickness and average 12 to 25 μ in diameter and from 25 to 30 mm. in length. The cells are regularly built up, cylindrical in shape with nodes arranged at regular intervals. The nodes are stained more decidedly with methyl violet or more deeply when treated with zinc iodochloride. The cell wall is of constant thickness and leaves but a narrow internal channel, which therefore appears merely as a dark line and is in many cases undistinguishable. The spun fibres are mostly folded longitudinally, and the natural ends are sharp pointed and generally long drawn out. A characteristic appearance is afforded by the cross section (see Fig. 14), which exhibits a number of loosely joined, acute-angled polygons, without any yellow circumferential stain when treated with sulphuric acid, the lumen showing as a yellow spot through the protoplasm. The fibre is free from woody fibre, and swells up in ammoniacal copper oxide without dissolving completely therein.

Microchemical Reactions.—Iodine and sulphuric acid give a blue stain, less quickly developed than in the case of cotton; madder tincture, orange; fuchsine, followed by a little ammonia, gives a permanent rose coloration (distinction from cotton); caustic soda, faint yellow stain.

Other means of detecting cotton in linen goods are afforded by :—

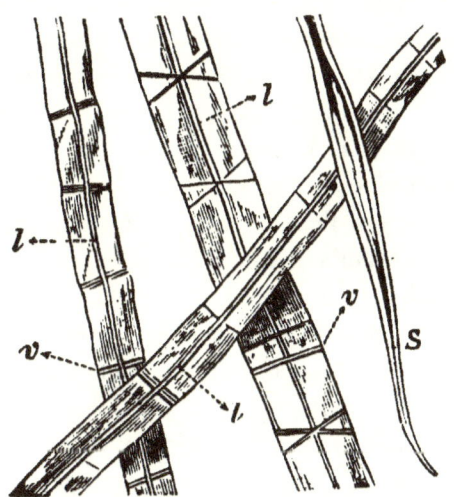

13. Magnified flax fibres: *l*, lumen; *s*, apex; *v*, nodes.

Fig. 14. Cross section of flax fibres.

(*a*) Treating the mixture of cotton and linen with a solution of caustic potash (1 : 6). The flax will become more curly than the cotton, and the latter finally turns greyish white, whereas the flax is dyed orange colour (Kuhlmann method).

(*b*) Treating the mixture with a stronger solution of caustic potash (1 : 2) by boiling for two minutes, then washing, and drying between blotting paper. Flax becomes a deep yellow, cotton whitish or straw colour (Böttger's method).

(c) The mixture is boiled in water and then steeped in concentrated sulphuric acid for two minutes. Cotton is dissolved but linen remains white and unaltered, and can be separated by washing with a weak solution of caustic potash.

(d) The mixture is boiled in water, dried, dipped for a few moments in thin, clear oil or glycerine and then pressed. Linen, by its greater capillarity, becomes transparent, whilst cotton remains opaque (Simon's method).

(e) The mixture is boiled in water, dried and dipped in a strong solution of common salt and sugar, the fibres being then burned over a flame. Linen leaves a grey, cotton a black, ash (Chevalier's method).

(f) The mixture is dipped in an alcoholic 1 % solution of fuchsine, washed and then laid for three minutes in ammonia. Cotton remains uncoloured, but flax is dyed rose-red (Böttger's method).

(g) The mixture is dyed in an alcoholic extract of cochineal (or madder root). Cotton becomes pale red (or yellow), but flax is dyed violet (or orange, or red) (Bolley's method).

(h) Cotton threads when viewed by the eye appear of regular form throughout, whereas flax threads are irregular. When quickly torn across, cotton threads curl up, but flax threads remain smooth : an accurate judgment by this means is, however, acquired only after long practice.

(c) Hemp.

The bast fibres of the stem of the hemp plant (Cannabis sativa) are mostly from 15 to 25 mm. long, whilst the diameter varies from 16 to 25 μ. As is shown by Fig. 16 the cells are very irregular in form, being partly band-shaped and partly cylindrical. Carefully isolated cells exhibit no parallel striations, but these are more or less decidedly apparent in the spun fibre, especially when stained with methyl violet. The lumen is mostly broad and becomes linear towards the extremity of the fibre, but the transformation is not sudden. The cell wall is not of such constant thickness as is the case

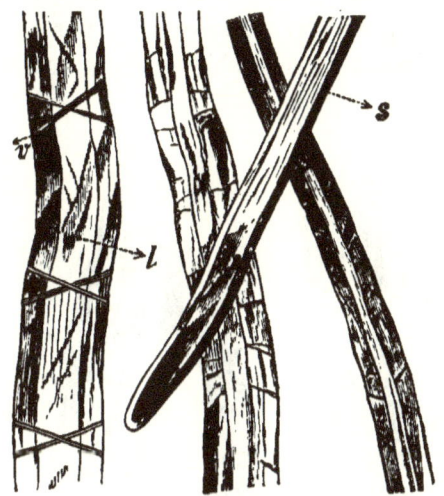

FIG. 15. Magnified fibres of hemp.

FIG. 16. Cross section of hemp fibres.

with flax. The ends of the fibres are blunt (distinction from flax), very thick-walled, and frequently branched laterally; cross stripes are frequently met with, but no nodes. The cross section shows the fibres attached in dense groups; the edges are generally rounded and exhibit a yellow circumferential stain when heated with iodine and sulphuric acid.

The lumen, examined in cross section, is not a mere dot like in flax, but is linear, frequently branched and irregular, without any contents.

Microchemical Reactions.—Iodine and sulphuric acid give a bluish green or dirty yellow coloration; aniline sulphate, a more or less yellow colour (faint lignification); hydrochloric acid, brown; caustic potash, brown; ammonia, faint violet; sulphuric acid gradually dissolves the fibres; ammoniacal copper oxide causes considerable swelling and effects partial solution.

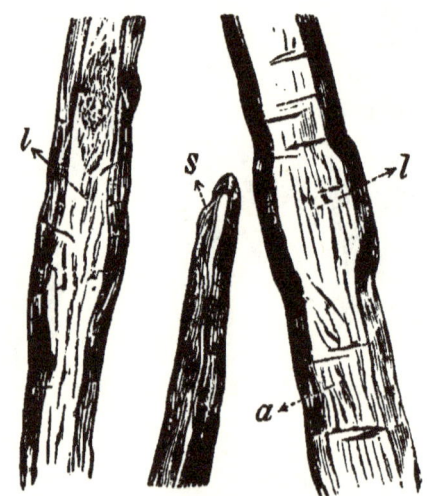

FIG. 17. Magnified nettle fibres: a, fissures in the wall; l, lumen.

(*d*) NETTLE FIBRE, CHINA GRASS (Ramie).

In recent times the bast fibres of the stems of various foreign nettles (Urtica) have been extensively employed. These fibres are generally 120 mm. long, the diameter ranging from 25 to 110 μ. When subjected to mechanical and chemical preparation the fibre is snow-white and highly lustrous, but unfortunately loses this property in the spinning process as at present practised. The cells are partly cylindrical, partly tubular; more rarely wide or band-shaped. The lumen is broad and mostly measures ½ to ⅔ of the entire diameter, sometimes less but rarely more. Frequently, lines are observed stretching across the cells, and a granular protoplasm is usually discernible. The cell walls are of even thickness, so that the lumen is of regular dimensions; the extremities terminate in thick-walled rounded points and have a linear lumen. In cross section the fibres are seen to be always isolated, very large, usually elongated and com-

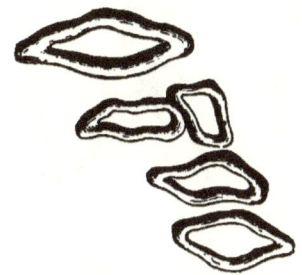

FIG. 18. Cross section of nettle fibres.

pressed flat, with, however, an open lumen frequently showing granular contents.

Microchemical Reactions.—Iodine and sulphuric acid stain blue (pure cellulose). Ammoniacal copper oxide causes great distensions but has no solvent action. Aniline sulphate induces no change.

(e) JUTE.

By the term jute is understood the bast fibres of the stems of several kinds of Corchorus, especially C. capsularis and C. olitorius. The crude fibre has a fine lustre and is some 140 inches long, the diameter varying between 10 and 30 μ. It is usually whitish yellow in colour but turns brown after prolonged storage. The cells exhibit a somewhat remarkable structure, owing to the irregular thickness of the cell wall, in consequence of which the internal and exterior border lines are not parallel, so that the lumen is alternately enlarged and contracted to a faint line. In commerce, however, some jute fibres are encountered which do not exhibit this variation in any remarkable degree, but a frequent interruption of the lumen is observed. The cell wall always exhibits considerable powers of refraction and therefore seems

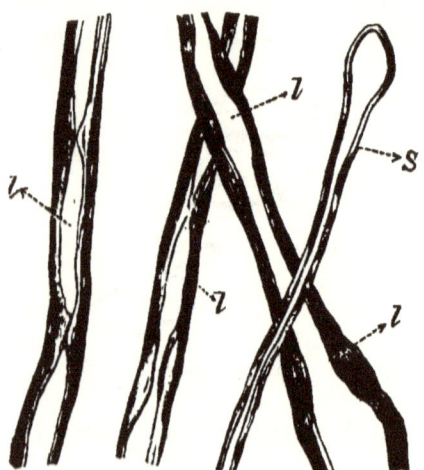

Fig. 19. Magnified jute fibres.

Fig. 20. Cross section of jute fibres.

to be very greatly limited by the internal space. The low tensile strength of jute, and its tendency to ravel, are attributed to the variable thickness of the cell walls.

Microchemical Reactions.—When treated with dilute chromic acid to which a little hydrochloric acid has been added jute turns blue; iodine and sulphuric acid give a dark yellow stain; aniline sulphate, a strong yellow coloration (therefore lignification of the fibre); ammoniacal copper oxide causes distension.

To distinguish jute from flax and hemp the threads are warmed in a solution of nitric acid and a little potassium chromate, washed, warmed in alkaline water and washed again; when the water is evaporated from the slide a drop of glycerine is added, and after a short time the characteristic structure of the jute will be definable. Under the polariscope (crossed Nicol prisms and dark field of vision) jute fibre shows a uniform blue or yellow colour, whereas linen or hemp is beautifully prismatic. The use of phoroglucine chloride is also advisable as a distinguishing test. Moistened with this reagent, jute stains an intense red, flax remains uncoloured and hemp is dyed a somewhat reddish tinge.

This terminates the description of the chief fibres extensively employed in the textile industry, but there still remain a few which are partly spun into yarn for upholstery goods, etc., but for the most part find employment in the manufacture of rope.

Manila Hemp.—The bast fibre of Musa textilis and other varieties comes from the East Indies and forms the best material for ropemaking. The light coloured fibres are hackled and spun into yarns which are used for making market-bags and similar weavings. Latterly the finer sorts are also used as weft for coarser upholstery goods. The fibre is from 60 to 280 inches in length, highly lustrous, smooth and even. The diameter ranges between 16 and 27 μ, of which $\frac{1}{4}$ to $\frac{1}{2}$, and not infrequently as much as $\frac{3}{4}$ to $\frac{4}{5}$, is occupied by the lumen. The cell is of regular structure, moderately thick and tapers gradually to a point.

Microchemical Reactions.—Iodine and sulphuric acid, golden yellow stain; caustic soda, faint yellow with slight distension; ammoniacal copper oxide produces considerable distension, but does not dissolve the fibre.

Manila hemp can be distinguished from Sisal by the colour of the ash, that of the former being greyish black, whereas Sisal leaves a white, and a mixture of the two sorts a greyish white and black ash.

New Zealand Flax.—The bast fibres from the leaves of Phormium tenax are worked up in New Zealand into yarn and cloth, or used in the crude state for making cord and rope. The fibres vary in diameter between 8 and 29 μ, the lumen constituting $\frac{1}{4}$ to $\frac{1}{3}$ of this width and being of regular dimensions; the ends are sharply pointed and the cross section circular.

Cocoanut Fibre (Coir).—The red-brown fibres constitute the bundles of bast tissue surrounding the hard shell of the cocoanut, and are chiefly used for making yarn mats and ropes. The bast cells are very short, being only $\frac{1}{2}$ to 1 *mm.* in length; the diameter increases regularly towards the centre, and so, in concordance therewith, does the breadth of the lumen, which occupies $\frac{1}{3}$ to $\frac{2}{3}$ of the width of the cell, according as the cell walls are thicker or more slender. At about the centre of the cell the inspissated layers draw together, so that the internal hollow is divided. Pores are numerous. The fibre is from 12 to 20 μ in diameter, and round or oval in cross sections, the latter being of a yellow-brown colour.

Cosmos Fibre.—This product (first manufactured near Brussels), which appears from time to time under various names, and has been recommended as a substitute for cotton, wool and silk, consists of manufactured residues from linen, hemp and jute. It is most frequently spun in conjunction with wool.

2. WOOL.

There are but few animals whose hairy covering finds employment in the textile industry. The largest quantity is obtained from the sheep, the yield from the Angora goat (mohair) and the alpaca playing a subordinate part. The dimensions of the wool hairs vary not only in different animals, but also in the different parts of the body of the individual.

Generally, hair is tubular, frequently curly, containing an internal medullary substance, and presenting a more or less scaly appearance on the outside. Underneath the scales is a layer of fibrous texture, frequently very faint, so that it is difficult to detect, especially when the medulla is strongly developed.

The sheep produces two kinds of hair, the very curly wool and a sleek hair (beard hair), but it is only in the ordinary native sheep and a few rare varieties of goat that both classes are present together—a long coarse upper hair and a much finer and shorter body hair (down). Otherwise the body is covered entirely, either with curly or smooth hair alone. The fineness of the hair in the higher races of sheep varies in different parts of the body, but only the curly hair is endowed with the property of felting.

In the examination of wool these differences must be borne in mind. The wool consists either—

(a) *Of pure curly wool*, from the merino and allied races of sheep, such as the South Down, Hampshire Down, Electoral and Negrettir sheep ;

(b) *Of pure sleek hair*, e.g., from the English Cheviot (Leicester breed) ; or

(c) *Of a mixture of both kinds*: ordinary native wools, such as native Australian, German, Russian, South American, etc., wool.

In cross section, wool is mostly circular, whereas that of

the hairs of furred animals presents the appearance of a flattened ellipse.

As already indicated, three layers are distinguishable in the fibre of wool.

(1) The upper layer of scales (particularly characteristic of sheep's wool), overlapping like roofing tiles, and arranged with more or less regularity, according to the fineness of the wool, surrounding the circumference to a greater or less extent, and finally either plain or exhibiting longitudinal striations, these being especially visible in wool containing no medulla. The scales show up plainly when the wool is treated with ammoniacal copper oxide or chromic acid.

(2) Underneath the scales rests the true fibrous material, generally colourless but sometimes also coloured, which, in the case of shoddies—from which the scales have been worn by previous treatment—makes its appearance on the surface and is then characteristic (see later).

(3) The medullary matter which fills the central portion of the tube and, especially in beard hairs, appears in more or less insular masses, but is lacking in the finer grades of wool. Under the microscope this matter is dark, but may be rendered transparent by boiling the fibre in glycerine and oil of turpentine.

Chemically considered, wool consists of keratin (horny matter), but also contains, in addition to carbon, oxygen and hydrogen, about 17 % of nitrogen and 5 % of sulphur. This composition is strikingly revealed, on combustion, by the peculiar unpleasant smell of horn, which distinguishes wool from cotton.

On this account the chemical reagents employed in testing wool differ considerably from those used for vegetable fibres. Boiling caustic potash or soda dissolves the fibre very readily; if acetic acid be added to this solution, sulphuretted hydrogen is evolved and a precipitate formed. Concentrated sulphuric acid leaves the fibre unaltered a short time, merely loosening the scales, but dissolves it fairly quickly to a reddish-brown solution on boiling; and hydrochloric and nitric acids also dissolve wool.

(a) Sheep's Wool.

This wool is generally white or yellow, rarely black, in colour, from $1\frac{1}{2}$ to 12 inches in length and between 14 and 60

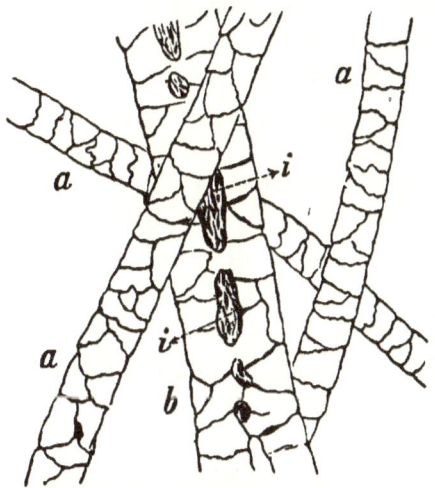

Fig. 21. Magnified fibres of wool: *a*, merino; *b*, Leicester (beard hair); *i*, medullary cells.

μ in diameter, the length and fineness of staple and the form and number of the curls forming the basis of commercial classification. Sheep's wool is distinguished from all others by the quantity of scales, and these are more strongly and regularly developed in proportion to the fineness of the wool. The outer circumference is generally toothed. In the finer wools the edge of the scales extends over the whole width of the hair, and the medullary substance is lacking, whilst in coarse wools the scales are small and irregularly placed, and the medulla appears in the form of long or rounded islands.

The wool hairs are of equal diameter throughout their entire length, lamb's wool alone tapering off gradually to a point.

In tanner's wool, obtained by un-hairing fells, as also in glover's wool and all wool from dead sheep, we usually find a mixture of various classes of wool, frequently containing particles of lime as an impurity. Very often the roots of the hair are present, and are readily distinguishable by their oval form; the wool itself is particularly brittle and is therefore only suitable for spinning along with sound wool.

Microchemical Reactions.—Ammoniacal copper oxide produces considerable distension and brings the scales into view. Concentrated hydrochloric acid and sulphuric acid gradually

FIG. 22. Cross section of wool fibres.

dissolve the wool, with red coloration; nitric acid dissolves it with difficulty, producing a yellow coloration; and cupric or ferric sulphate dyes the wool black.

(b) MOHAIR WOOL.

The hair of the Angora goat has a silky lustre, is extremely fine and smooth, of considerable length, and but slightly, if at all, curly. The scales, which are regular and surround the entire hair, are only discernible under very close examination. The length of the wool is from 4 to 6 inches, and the width 26 to 30 μ. The fibres appear to be free from medullary substance, except in very thick individual hairs, where it shows up decidedly in the form of a central canal occupying about one-half the entire diameter. This wool is characterised by fine, regularly arranged fissures in the fibre, which also appear as fine dots on the surface.

(c) Alpaca Wool.

The long, soft, silky hair is naturally white, grey, brown or black, 4 to 6 inches long and 20 to 34 μ wide. The lustre is inferior to mohair, the scales extremely fine, or more often absent, whilst, on the other hand, longitudinal lines and small elongated islets of medullary substance are visible. This fine wool should not be confounded with alpaca shoddy.

The following kinds of hair are of minor importance:—

Cashmere or Thibet Wool.—The hair of the Cashmere goat consists of a fine soft down and a sleeker, long, beard hair, the former exhibiting scales without any medullary substance, whilst the latter is very decided in the beard hairs. The down is $1\frac{1}{4}$ to $3\frac{1}{2}$ inches long and 13 μ wide, and the hair $3\frac{1}{2}$ to $4\frac{1}{2}$ inches by 60 to 90 μ.

Vicuña Wool.—The pure vicuña wool (which should not be confounded with the artificial product of the same name) is obtained from South America, and is rarely met with in commerce. It is very soft and delicate, of a reddish-brown colour, and resembles alpaca wool. In this case also there are two classes of hair: the fine under hair and the coarse upper or beard hair, the first-named being covered with regular scales and generally free from medulla, whilst in the latter the medullary substance is strongly developed and dark in colour, frequently divided by lighter central stripes. The edges of the scales are much less distinct than in sheep's wool. The under hair is from 10 to 20, and the upper hair about 75 μ wide.

Llama Wool.—The upper hair of the llama also exhibits a strongly developed insular medullary substance without any scales being detectable. In the under hair the scales are more faintly developed than in vicuña wool, but both upper and under hairs show longitudinal furrows. The former is 150 and the latter between 20 and 35 μ wide.

Camel Hair.—The camel yields a very fine reddish or yellow-brown under hair, known in commerce as camel wool, and is used, among other purposes, for making Jäger's normal cloths, the coarser hairs being employed for making carpets, coverlets, etc. Both upper and under hair exhibit faint scales, but strongly developed longitudinal

FIG. 23. Camel wool.

furrows. The medullary substance always appears in the upper hair, but not so decidedly as in the case of vicuña and llama hair, whereas in the under hair it is either wanting or appears, though rarely, in insulated masses. The under hair is 14 to 28 μ wide, and the upper hair 75 μ wide.

Hare and Rabbit Fur.—The hairs are brown to black in colour, $\frac{1}{2}$ to 1 inch long, and 76 to 100 μ wide, tapering out to a single fine point. The medullary canal is very regular and composed of several rows of cells, so that this hair can readily be distinguished from all others under the microscope.

Horse Hair.—This consists almost exclusively of mane and tail hairs, chiefly white and black. The hair is very long and ranges in diameter from 90 to 250 μ, the strongly developed medullary canal being highly characteristic.

Cows and Calves' Hair.—Cows' hair from Siberia (chiefly) is used now-a-days in the production of carpet yarns. The colours are preferably white, reddish or black, and with a dull lustre. The medullary canal occupies $\frac{1}{2}$ to $\frac{3}{4}$ of the total diameter.

3. ARTIFICIAL WOOLS.

The "artificial" wools comprise the products elaborated from old or new wool waste, or recovered from rags, and are generally divided into three classes:—

1. *Shoddy* is the wool recovered from old long-staple materials, such as stockings and other knitted goods, and is spun by itself as shoddy yarn.

2. *Mungo* is a short-staple article from milled goods, chiefly fragments of cloth, and can only be spun into yarn when mixed with longer wool or with cotton.

3. *Extract wool or Alpaca* is the name given to wool recovered from worsted rags by a chemical process known as carbonising. The fibre is mostly of short staple, and, as a result of defective manipulation, frequently has a corroded appearance when examined under the microscope. The carbonising process is effected by means of sulphuric acid, which destroys the cotton present.

Whilst the foregoing particulars will facilitate the microscopic differentiation of all kinds of vegetable fibres, as well as silk, from wool, it is not such an easy matter to determine whether the wool in yarn or cloth is a pure natural wool or artificial.

Testing.—A thread is prepared and examined under the microscope by a low power; this will enable one to detect wool and silk or cotton and linen if present side by side. By now adding a drop of ammoniacal copper oxide, the silk and cotton will be immediately dissolved, the linen more slowly, and finally the wool will become slightly distended. For the second test, concentrated sulphuric acid is added, whereupon

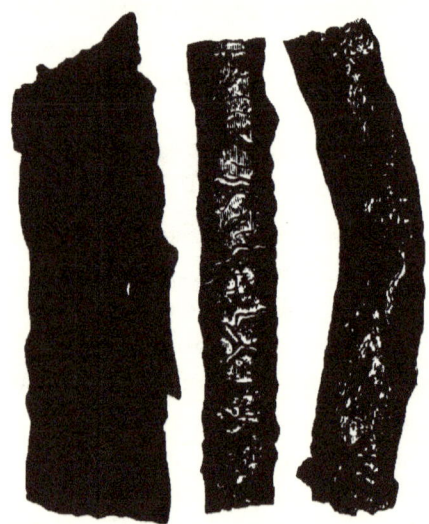

Fig. 24. Shoddies.

the wool is dissolved with a reddish coloration, its composition being finally ascertained by a third test.

The chief characteristic test for artificial wool is the colour, which is seldom uniform under the microscope, even pure white, red or green yarns always containing fibres of different colour. Generally, coloured and plain fibres are seen side by side, the latter being either pure white or retaining a trace of their original colour. In the better kinds the dyed hairs are uniform, but multicoloured in the inferior sorts. In this examination it is advisable to warm the sample up beforehand with hydrochloric acid, which will remove the colour due to the second dyeing and leave the original dye clearly exposed.

A further confirmatory test is afforded by the ends of the fibres, these being usually unbroken in the case of natural wool, whereas in the artificial varieties they are always torn and ragged, the scales and medulla breaking off clean and leaving the fibrous layer in the form of a brush.

Another confirmation of the presence of artificial wool is given by the absence of, or at any rate the corrosion of, scales, only traces of their existence being discoverable at the edges; finally artificial wool is never so uniform in diameter throughout the length of the hair as natural wool, the fibre narrowing and expanding again abruptly.

Professor Dr. Von Höhnel in his admirable work on the microscope structure of the textile fibres [1] gives the following as the chief factors in the detection of artificial wools:—

(1) *Extraneous Fibres.*—Pure textile fabrics always consist of a single kind of fibre. In no case should a refuse wool hair be found in conjunction with merino hair; nevertheless it may happen that stubby hairs appear, though this is no proof of an intentional admixture of inferior hair. Should several per cents. of dyed cotton be found in a material, that is a sure indication of the presence of artificial wool, for it seldom occurs that good wool is adulterated direct with cotton. Undyed cotton, unless present in any remarkable quantity, need not give rise to suspicion.

(2) The length of fibre is only in individual cases proof of the presence of artificial wool; *e.g.*, in worsteds where the entire length of the fibre can be revealed by carefully teasing out the thread, an operation almost impossible of performance

---

[1] *Lehrbuch der Mikroskopie der Gespinnstfasern*, p. 109.

in the case of milled cloths. The addition of very short fibres can be detected by brushing the sample with a stiff brush, the percentage of short waste fibres from both sides of good cloth not exceeding ½ per cent. If more is produced the cloth should be more accurately examined, in which case it is advisable to specially test the waste fibres.

(3) The thickness of the hair is a very unreliable indication; the more uniform the diameter of the threads in a

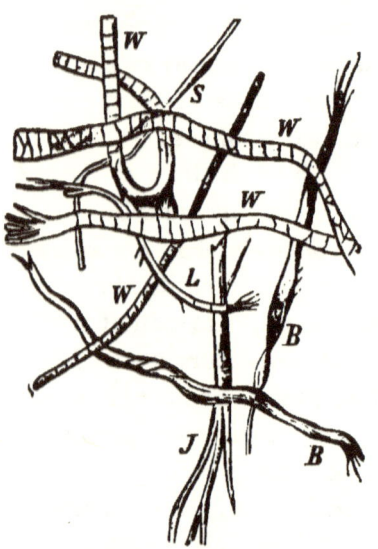

FIG. 25. Shoddy yarns (magnified seventy-fold): W, wool; B, cotton; L, linen; S, silk; J, jute.

woollen cloth the better will the latter be; but only when the differences of thickness are remarkably large can the presence of shoddy be suspected.

(4) *Histological Properties of Artificial Wool.*—Good wool almost invariably exhibits decided scales. On the other hand inferior wool, even on the living sheep, is deficient in scales from the tip downwards, though of otherwise normal structure. The absence of scales cannot therefore by itself

sufficiently prove the presence of artificial wool, though in fine wools they should never be lacking: so that if a merino or other fine wool be found deficient in scales this will indicate the presence of shoddies, whereas in coarse wool the circumstance would not arouse suspicion. Very curly merino wool is apparently not so easily stripped of its scales by rubbing, etc., as is the stout, stiff common wool, the individual hairs of which stand out separately from the body of the sheep.

(5) The ends of the shoddy fibres, being of different form to those of natural wool, present a sure means of identification. In working up artificial wool the threads are always torn and the epidermis stripped clean off along with the medullary matter, leaving the fibrous layer projecting like a brush, especially when swollen by the aid of hydrochloric acid. If many or the most visible ends are torn shoddy is indicated (or mungo if the fibres are short).

(6) The colour is an equally important proof. Many parcels of rags are of one single colour, but for the most part they are composed of variously coloured wools, the result being that but very few shoddy samples are uniform in colour, even apparently pure white, red, yellow, or green threads actually containing fibres of other colours. Therefore if, in a yarn of any particular colour, there are found a number of individual fibres of variable, and generally very glaring colours, the presence of shoddies can be assumed with certainty.

General Remarks.—When a fabric or yarn is of a brownish or blackish grey, and, above all, dirty colour, and is composed of threads of all colours, it will for the most part consist of shoddy. If a material contains sheep's wool and, mostly, dyed cotton, speckled or twisted together, or contains a whole or semi-cotton warp, a large amount of shoddy will be present. Woollen goods containing vegetable fibres are rarely made from natural wool. Shoddy yarns are, especially in winter goods, found in the under weft at the reverse side of the cloth. Such threads are generally thicker, more tightly twisted, and curlier (less smooth) than yarns from pure wool. Frequently a thick shoddy yarn is found twisted with a thin strong wool yarn.

In completion of the examination of artificial wools the following tests, in addition to the microscopical, are now given :—

CHEMICAL EXAMINATION.

(a) Water Determination.

Like all other fibres, those of artificial wool are hygroscopic. The estimation of their water content is effected by weighing a sample in a weighing bottle and re-weighing after drying at 100° C., the operation being repeated until the weight is constant. The difference between the initial and final weights indicates the amount of water present.

(b) Estimation of Fat.

The dried sample is freed from fat by extraction with petroleum spirit, the fat being recovered by carefully distilling the solvent, and weighed.

(c) Estimating the Percentage of Cotton.

The sample from (b) is boiled for a quarter of an hour in 8° B. soda solution, whereby the wool is dissolved, leaving the cotton fibre unchanged. This residue is collected on a linen filter, and after washing with hot water until the alkaline reaction disappears, is dried and weighed. If the artificial wool contains many impurities, the sample must, before performing this estimation, be washed with slightly acidified water, and afterwards with pure water, to remove the greater portion of the impurities which, otherwise, would have been weighed as cotton. This estimation is, however, merely an approximate one.

(d) Determination of the Ash.

This estimation is required but seldom, *e.g.*, when it has to be determined whether the waste from cloth in process of manufacture proceeds from the shoddy used or not. In such event by estimating the ash of the waste and that of a sample of the shoddy the qualitative equality or otherwise of the results will solve the question. The operation is performed in a porcelain crucible, the sample being slowly incinerated until carbonisation ceases, and then heated to redness by the aid of the blow-pipe flame. The qualitative analysis of the residue is then performed in the usual manner.

4. SILK.

Silk is the fibre with which the larvæ of various insects surround themselves before entering the pupal stage, the envelope being termed a cocoon. The cocoon thread results from the hardening of the fluid ejected from the two serecteria of the larva; hence it follows that silk exhibits no definite structure, but consists of cylindrical, sometimes flattened, or, more rarely, helical (twisted round the axis) compact threads. The colour is white, pale or dark yellow, or occasionally reddish, the colouring matter being contained in the external layer, so that when this is removed the silk is white.

The different kinds of silk are distinguishable by their diametrical dimensions, and by the more or less decided appearance of longitudinal stripes, or by polarisation colours. In examining raw silk it is advisable to first boil the sample in a solution of soap. Chemically considered, silk is composed of three chief substances:—

(1) Silk gelatin or sericin (22·5 per cent.).
(2) True silk fibre or fibroin (63 per cent.).
(3) Fat, resin, colouring matter, and mineral substances (about 2 per cent.).

FIG. 26. Silk fibres.

FIG. 27. Organzine silk.

Under the microscope silk appears as a smooth (more rarely striped) cylinder without any contents; in contradistinction to vegetable and wool fibres both lumen and scales are lacking. Mostly the silk appears as a double thread.

Microchemical Reactions.—The fibre dissolves with difficulty in soda and potash solutions; sugar and sulphuric acid dissolve it with a rose-red coloration (albumin reaction), hydrochloric acid with violet coloration. It burns in the same manner as wool, but since the fibre contains no sulphur, no smell of horn is evolved: otherwise it contains all the other elements present in wool, *viz.*, carbon, hydrogen, oxygen and nitrogen.

SILK FROM BOMBYX MORI.

This insect produces the most valuable silk and the greatest bulk of the so-called "true silks". The raw fibre is either white or yellow in colour, exhibits no structure, and is rarely striped. When treated with dilute chromic acid, the thread assumes a fine fibrous structure. The diameter of the fibres is at most 18 μ. Polarisation colours are very clearly exhibited.

CHAPPE SILK.

This product, also known as "spun" silk, from silkspinner's waste and from inferior cocoons, is difficult to distinguish from pure silk under the microscope. According to Höhnel, it can be differentiated, with more or less accuracy, by the irregular form of the thread, and especially by the remarkable irregularity of the surrounding envelope of sericin or gum.

TUSSAH SILK,

also known as "wild" silk, is grey to brown in colour, and has a vitreous lustre. The fibres appear as double threads, not, however, structureless, like true silk, but composed of a bundle of very fine fibres (fibrillæ), manifested by parallel longitudinal stripes. Polarisation colours are but slightly

FIG. 28. Tussah silk.

apparent; the diameter amounts to 52 μ maximum, the cross section is an elongated quadrilateral, and a number of fine and dense granules are displayed therein.

LOADED SILK.

The subjoined illustration of loaded silks is very instructive, showing, as it does, that the microscope can be of good

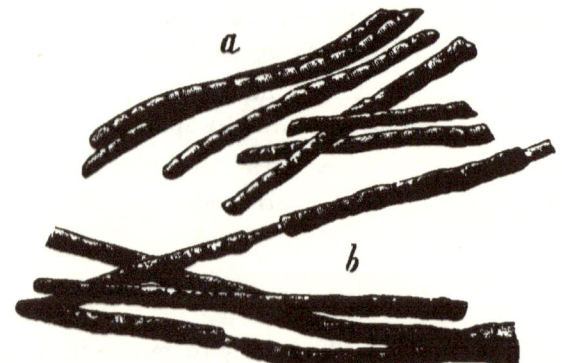

FIG. 29. Loaded silk: *a*, with 160/180 %; *b*, with 350/400 % loading.

service in detecting either low or high loading. In the latter case the thread appears to be entirely surrounded by the loading material, the rind being thicker than the fibre itself. In the case of lesser loading, it can be seen that the colour is taken up by the fibre.

Artificial Silk.

In view of the considerably lower value of the artificial silks, such as are offered by Chardonnet, it becomes important that we should possess means of identifying these also. The most suitable indications are the chemical and physical characteristics ; for example, the inferior strength and elasticity of artificial silks, and their deficiency in the " crackling " feeling observable in handling true silk. When dissolved in caustic potash they produce a yellow (pure silk a colourless) solution, and they are insoluble in an alkaline solution of copper containing glycerine.[1] These properties render the quantitative separation of artificial and natural silks possible.

II. CHEMICAL EXAMINATION OF TEXTILE FIBRES.

The textile chemist has not always a microscope at his disposal, and he therefore has frequently to rely on chemical reactions alone. Occasionally the imitation in the laboratory of manufacturing processes (*e.g.*, carbonisation) gives inaccurate conclusions respecting the presence of vegetable fibres in wool. The dyer knows that vegetable fibres are not coloured by the so-called acid dyes used in wool dyeing, whereas other colouring matters such as the Mikado dyes

---

[1] Prepared by dissolving 10 grms. of cupric sulphate in 100 grms. of water, adding 5 grms. of pure glycerine and then sufficient caustic potash to redissolve the precipitate first formed.

(Leonhardt) act in a converse manner. For testing for cotton and linen in a mixed cloth employment is made of fuchsine solution, followed by steeping in ammonia, which destroys the colour in the cotton fibres but leaves the linen fibres dyed a pale rose colour. By using fuchsine solution prepared according to Liebermann's directions, wool can be detected by the rose coloration in a mixture of wool and vegetable fibre. Silk and wool are separated by means of boiling hydrochloric acid, which dissolves the former, whereas the latter merely swells up without passing into solution. Persoz observed that silk dissolves readily in a 60° B. solution of basic zinc chloride, and on this circumstance a quantitative method was based by Renouard.

A very simple and often sufficient means of distinction between wool thread and vegetable fibre consists in a combustion test, the nitrogenous and sulphurous wool fibre evolving an odour of burnt horn and leaving behind a characteristic nodule of ash, the operation being one of carbonisation rather than true combustion. On the other hand, cotton fibre burns away very quickly without the slightest smell. Further chemical examination must, however, be made to ascertain whether the wool is pure or mixed with cotton, and for this purpose strong solutions of alkalis, which dissolve animal, but not vegetable, fibres, are used. The mineral acids, such as hydrochloric and sulphuric acids, behave conversely by dissolving vegetable fibres when heated but scarcely corroding those of animal origin.

The preparation of the test solutions for examining textile fibres both qualitatively and quantitatively will now be described.

(1) *Ammoniacal Copper Oxide Solution.* (a) *Schweitzer's Method* (1857).—By carefully precipitating cupric dithionate with dilute ammonia, pale green basic cupric dithionate is obtained, and after filtering and washing is redissolved by

warming in ammonia. On cooling, crystals of ammonium dithionate separate out, leaving ammoniacal copper oxide in solution in the supernatant liquid.

(b) *Böttcher's Method.*—A glass tube, 24 × 2 inches, is loosely filled with thin rolled copper, and after being fitted at the lower end with a pinchcock, is filled with ammonia, the liquid being drawn off into a glass after a few minutes and then poured over the copper again. By proceeding in this way for several hours a deep blue liquid, thoroughly saturated with cupric oxide, is obtained.

(c) *Neubauer's Method.*—A solution of cupric sulphate is precipitated by caustic soda in presence of sal ammoniac, the precipitate obtained being, after decantation and washing on a filter, stored under water. The solution is prepared by adding this precipitate to an excess of ammonia so long as the latter will dissolve it, a deep blue solution being thus obtained.

(d) *Wiesner's Method.*—By covering copper turnings with a 13 to 16 per cent. solution of ammonia in an open flask.

Application.— The solution dissolves cellulose (cotton) with ease, and causes lignified fibres such as hemp to swell up.

(2) *Sodium Copper Oxide Solution. Löwe's Method.*—Dissolve 16 grams cupric sulphate in 140 to 160 *c.c.* of water and add 8 to 10 *c.c.* of glycerine (sp. gr. 1·24), mixing thoroughly by agitation. Then add by drops, taking care to avoid excess, sufficient cold caustic soda to redissolve the light blue precipitate of cupric hydrate at first formed. The filtered ultramarine blue solution will keep for a long time if properly stoppered.

Use.—Silk fibre dissolves slowly in the very dilute, but swells up quickly in a moderately concentrated solution, and dissolves to a thick liquid when more is added. Silk dyed black by salts of iron is imperfectly and less readily soluble. On the other hand if such silk be immersed for some time in a solution of potassium-or ammonium sulphide, washed, and the iron sulphide removed by dilute hydrochloric acid, the solution in the sodium copper sulphate is more readily effected, and the same object can be accomplished by a preliminary treatment with zinc and dilute hydrochloric acid.

(3) *Ammoniacal Nickel Solution.*—25 grams of crystallised nickel sulphate are dissolved in 500 *c.c.* of water, and nickelous hydrate thrown down by caustic soda. After washing, the precipitate is redissolved in 125 *c.c.* of concentrated ammonia and 125 *c.c.* of water.

Use.—This liquid dissolves silk immediately, but reduces the weight of linen and cotton by only 0·45 per cent. and pure wool by only 0·33 per cent.

(4) *Zinc Chloride Solution.*—According to Persoz a concentrated solution of sp. gr. 1·7 (= 60° B.) is used. Elsner prepares a basic solution by dissolving 1000 grams of dry zinc chloride in 850 *c.c.* of distilled water and adding 40 parts of zinc oxide. The sp. gr. of the solution is 1·65.

(5) *Iodine Solution.*—Höhnel's recipe prescribes dissolving 1 gram of potassium iodide in 1000 parts of water and adding thereto iodine in slight excess. A more concentrated solution is prepared by dissolving 3 grams of crystallised potassium iodide in 60 *c.c.* of water and adding 1 gram of iodine, the resulting dark brownish red liquid being diluted, as required, by the addition of distilled water. It is advisable to prepare the solution shortly before use.

The simplest way to bring about the iodine reaction in preparations under a cover glass in water is by introducing a fragment of iodine.

The sulphuric acid employed in connection with this reaction consists of 3 parts of concentrated sulphuric acid, 1 part of water and 3 of concentrated glycerine.

CHEMICAL EXAMINATION OF TEXTILE FIBRES. 47

(6) *Fuchsine Solution. Liebermann's Method.*—Caustic soda or potash is added drop by drop to a saturated solution of fuchsine until the latter is decolorised, the filtered solution being then stored in a well stoppered bottle.

Use.—For distinguishing between undyed animal fibre and vegetable fibre. If the mixed fibre be immersed in the hot solution wool and silk will be found, on rinsing, to be dyed red, the vegetable fibres remaining uncoloured.

(7) *Aniline Sulphate Solution.*—A concentrated aqueous solution of the salt, acidified with a little sulphuric acid, is used; or the chloride may be employed, in which event hydrochloric acid must be added in place of sulphuric acid.

Use.—For detecting woody tissue.

(8) *Phloroglucin Solution.*—3 grams of phloroglucin are dissolved in 25 *c.c.* of alcohol and mixed with 25 *c.c.* of concentrated hydrochloric acid.

Use.—For detecting jute fibres, which it dyes an intense red, hemp and flax being merely coloured a faint rose-red.

(9) *Naphthol Solution.*—20 parts a-naphthol dissolved in 100 parts of alcohol.

Use.—Characteristic colorations are produced in presence of vegetable fibres.

(10) *Mineral Acids.*—Nitric acid (commercial), sulphuric acid, 58° B. strength; hydrochloric acid, 3 per cent. strength.

(11) *Alkalis.*—Soda solution, 0·1 per cent.; potash solution, 10 per cent.; caustic soda solution, 7° B. (sp. gr. 1·05).

The behaviour of the various fibres under the action of the several acids, bases and salts is given in the subjoined tables :—

I. CHARACTERISTIC COLORATION BY DYE STUFFS.

	Wool.	Silk.	Flax.	Cotton.
Madder tincture	*nil*	*nil*	orange	yellow
Cochineal tincture (Decolorised by bleaching powder solution)	scarlet	scarlet	violet	light red
Fuchsine (Liebermann)	slightly red dyed	slightly red dyed	slowly *nil* *nil* dyed	rapidly *nil* *nil* dyed
Acid tar dyes				
Mikado dyes	—	—		

II. INFLUENCE OF VARIOUS SALTS IN SOLUTION.

	Wool.	Silk.	Flax.	Cotton.
Zinc chloride	partially dissolved	dissolved	fibre undissolved: yellow coloration	
Stannic chloride	no change	no change		coloured black
Silver nitrate	violet to blackish brown	no coloration		*nil*
Mercury nitrate (Millon's reagent)	brick red to brown	no coloration	*nil*	*nil*
Cupric or ferric sulphate	black	no coloration	*nil*	*nil*
Sodium plumbate (solution of lead in caustic soda)	black precipitate	no p.p.		*nil*
Ammoniacal copper oxide	swelling only	no change	Swells up and partly dissolves: blue coloration	
Ammoniacal nickel oxide	undissolved	dissolved		undissolved

III. INFLUENCE OF ALKALINE LIQUIDS, ETC.

	Wool.	Silk.	Flax.	Cotton.	Hemp.	Jute.
Caustic potash	dissolves	dissolves	swells up: fibre becomes brownish yellow, subsequently lighter	swells up: fibres coloured faint yellow only	fibres are coloured brown	
Caustic soda	dissolves gradually	dissolves gradually and becomes reddened	brown yellow	faintly yellow	brownish	
Ammonia	—	—	—	—	—	—
The addition of the following substances to the alkaline solution of the fibre:						
(a) Sodium nitro-cyanide produces	violet	no coloration	—	—	*unretted*: orange yellow; *retted*: faintly violet	—
(b) Lead acetate produces	blackening	—	—	—	—	—
(c) Cupric sulphate produces	violet, subsequently brown	violet	—	—	—	—

Woody tissue is detected in hemp, jute, etc., by the yellow coloration produced with aniline sulphate; rose red with indol and sulphuric acid; and orange with naphthylamine chloride.

IV. INFLUENCE OF ACIDS, METALLOIDS, ETC.

	Wool.	Silk.	Flax.	Cotton.	Hemp.	Jute.
Sulphuric acid	does not dissolve until heated	dissolves quickly in hot acid	dissolves quickly in cold concentrated acid	quickly dissolved	dissolves slowly	dissolves slowly
Nitric acid	colours yellow and dissolves slowly	colours yellow and dissolves quickly	dissolves without coloration	uncoloured	colours yellow	—
Chlorine water	becomes yellow and brittle	becomes yellower	bleaches	bleaches	yellow brown	turns violet on addition of ammonia
Iodine solution	—	—	yellow brown to yellow	yellow	—	light brown
Picric acid	yellow	yellow	—	—	—	—
Iodine and sulphuric acid (cellulose reaction)	—	—	swells up: blue coloration	same as flax	swells up gradually: greenish coloration	swells up gradually: yellow to brown coloration
Thymol and sulphuric acid (cellulose reaction)	—	—	red violet	red violet	—	—
Sugar and sulphuric acid (furfurol reaction)	rose red	rose red	—	—	—	—

CHEMICAL EXAMINATION OF TEXTILE FIBRES.

ANALYTICAL TABLE FOR EXAMINING A MIXTURE OF FIBRES. ARRANGED BY PINCHON.

The mixture is acted on by a 10 % solution of caustic potash or soda.

I. Part dissolves

Part dissolves
The dissolved portion is not blackened by lead acetate, but the insoluble part turns black with this reagent: **SILK AND WOOL.**

The alkaline solution does not blacken on addition of lead acetate: **SILK.**

None dissolves
The mass is blackened by lead acetate: **WOOL.**

II. Part remains undissolved

Zinc chloride solution is allowed to react.

None dissolves
Chlorine water (or ammonia) colours the fibre — Red brown: **NEW ZEALAND FLAX.**

The fibre is coloured red by fuming nitric acid:
- Alcoholic fuchsine solution colours the fibre — Not at all
  - Permanently — Caustic potash stains yellow — Iodine and sulphuric acid colour:
    - Yellow: **HEMP.**
    - Blue: **FLAX.**
  - Colour removed by washing — No yellow coloration produced by caustic potash: **COTTON.**

Part dissolves
On addition of lead acetate part is:
- Blackened — Not blackened
- Caustic potash partly dissolves the fibres insoluble in zinc chloride. The remaining fibres are soluble in ammoniacal copper oxide: **MIXTURE OF WOOL, SILK AND COTTON.**
- Picric acid produces partial yellow coloration, the residual portion remaining white: **SILK AND COTTON.**

III. The whole dissolves

Nitric acid produces partial yellow coloration, the rest remaining white: **MIXTURE OF FLAX AND COTTON.**

QUALITATIVE AND QUANTITATIVE ESTIMATIONS.

(a) *Detection of Vegetable Fibres in Presence of Animal Fibres.*[1] —The following method is based on the formation of sugar when cellulose is treated with acids. The sample must be thoroughly boiled with water in order to extract the "dressing". When this is done, about 0·1 gram is put in a test glass with 1 c.c. of water and two drops of an alcoholic solution (15-20 per cent.) of a-naphthol, and as much concentrated sulphuric acid as there is liquid already. Vegetable fibres, if present, are readily dissolved, and the liquid assumes a deep violet colour when agitated; wool or silk gives a more or less yellow to reddish-brown coloration. Thymol produces a beautiful red coloration.

(b) *Separation of Wool from Cotton.*—The dressing and colour being removed by boiling the sample in dilute hydrochloric acid, dilute lye, or by extraction with alcohol, ether, etc., a weighed portion is dried at 100° C., and placed in 4 parts of sulphuric acid and 1 part of water for twelve hours, then mixed with three volumes of absolute alcohol and water and filtered. The residue is washed in absolute alcohol until the washings are colourless, and afterwards with water, being finally dried and weighed to ascertain the weight of wool present.

(c) *Separation of Wool from Cotton.* — After freeing the sample from dye and dressing as before, and washing, a portion is dried and weighed, being then immersed in ammoniacal copper oxide for twenty minutes, after which water is added. The residue left after filtration is thoroughly washed, dried and weighed, the result giving the amount of wool in the mixture.

(d) *Separation of Cotton from Wool.*—The cleaned, dried and weighed sample is gently boiled for two hours in 8° B.

---

[1] Dingler's *Polyt. Journal*, vol. cclvi., p. 135.

caustic potash, then well washed and re-dried. During the boiling a few drops of water are added from time to time to prevent the alkali becoming too concentrated. After drying at 100° C. the residue is weighed, the result giving the weight of cotton and the loss that of the wool.

Instead of potash, 7° B. caustic soda may be used, boiling being restricted to a quarter of an hour.

(*e*) *Separation of Silk and Wool.*—These fibres may be separated by boiling in hydrochloric acid, in which the silk is readily soluble, whilst the wool merely swells up.

(*f*) *Separation of Silk, Cotton and Wool.*—After removing the dressing and dye, as already described, the sample is treated with ammoniacal nickel oxide, which dissolves the silk at once. The cotton is then dissolved out by means of ammoniacal copper oxide, leaving the wool behind. The treatment may also be varied by boiling the sample for two to three minutes in 1 per cent. hydrochloric acid, after the removal of the nickel solution, and then washing and weighing. The cotton and wool are separated by boiling in a 2 per cent. soda lye, the residue (wool) being rinsed, dried and weighed.

(*g*) *Separation of Silk, Cotton and Wool* (*Rémont's Method*).—Two samples of yarn, each weighing 2 grams, are dried, weighed and boiled for a quarter to half an hour in 200 *c.c.* of 3° B. hydrochloric acid, to remove the dressing, and are then thoroughly washed and pressed. One sample is immersed in a boiling solution of basic zinc chloride, for a very short time, then washed thoroughly, first in acidified, afterwards in clean, water, and dried. The loss in weight gives the amount of silk.

The second sample is boiled for fifteen minutes in 60 to 80 *c.c.* of caustic soda (sp. gr. 1·02), and then washed and dried, the loss in weight representing the proportion of wool. The residue is cotton, the dry weight of which must be augmented by about 5 per cent. to compensate for the corrosion of the fibre during the operation.

(*h*) *Separation of True Silk, Tussah Silk, Wool and Cotton.*—The mixture is at first acted on by boiling half a minute with concentrated hydrochloric acid, which immediately dissolves the true silk, the tussah silk being dissolved at the end of two minutes' further boiling. On treating the residue with hot caustic potash the wool is dissolved, and the cotton left behind by itself.

Höhnel[1] recommends the following method for distinguishing between true and wild silk: A saturated solution of chromic acid is diluted with an equal bulk of water, and if pure silk be immersed in this solution and boiled for a minute, the fibre will completely dissolve, whereas tussah silk is barely attacked at all, even when the boiling is prolonged to two or three minutes. Wool behaves similarly to true silk. A weak solution of zinc chloride—45° B., or sp. gr. 1·45—attacks true silk very rapidly, but acts on tussah silk only after longer exposure, so that this solution also may be used for distinguishing and separating them.

(*i*) *Detection of Flax and Cotton.*—The sample threads are dyed by immersion in alcoholic fuchsine solution (1 gram fuchsine in 100 *c.c.* alcohol), then washed with clean water until the colour ceases to run, and steeped in ammonia for about three minutes. The linen threads will be dyed rose colour, whereas the cotton threads will be decolorised.

For the purposes of quantitative separation the samples, previously freed from colour and dressing, by a suitable boiling in dilute hydrochloric acid or distilled water, followed by a thorough rinsing, are dipped for one and a half or two minutes in concentrated (66° B.) sulphuric acid, then rinsed out well, rubbed between the fingers and

---

[1] Höhnel, *Mikroskopie*, p. 150.

neutralised by steeping in dilute ammonia or sodium carbonate solution. After washing over again in water the threads are pressed between blotting-paper and dried, when linen threads will, as a rule, be found to have retained their structure whilst the cotton has dissolved after passing through a gelatinous stage in which it will tear like tinder.

QUANTITATIVE ESTIMATION OF THE LOADING OF SILK.

In testing silk the question of the nature and quantity of loading present has always to be solved, so the matter will now be briefly discussed. Following the directions given by E. Königs, manager of the Silk Conditioning Institution at Crefeld, the first thing to do is to determine the percentage of water in the silk; then the fat, by extraction with ether; and afterwards the gummy integument, by boiling in water. From the residue the Berlin blue is extracted by alkalis, reprecipitated by acids, and the filtered precipitate calcined with repeated additions of nitric acid. Each part of the ferric oxide so obtained corresponds to 1·5 parts of Berlin blue. Any stannic oxide present is also determined and calculated as stannic catechutannate, 3·33 parts of which are the equivalent of 1 part of oxide. The total amount of ferric oxide is likewise ascertained and deduction made of the quantity present in the form of Berlin blue and that (0·4 per cent., or 0·7 per cent. in raw silk) existing in the silk, the residue representing the ferric oxide in combination with the tannic acid from catechu or chestnut extract, 1 part of this oxide corresponding to 7·2 parts of ferric tannate. Should ferrous compounds of this nature be present the multiplying factor will be 5·1 instead of 7·2.

According to Moyret, loaded silk is examined as follows:—

(*a*) *Moisture Determination.*—In the absence of a conditioning apparatus it will be sufficient to dry 10 grams of silk on an oil bath at 120° to 130° C. and estimate the water by the ensuing loss in weight. If this loss exceeds 15 per cent. it may be assumed that the silk has been loaded with hygroscopic substances.

(b) *Soluble Loadings.*—The dried silk is boiled in distilled water, rinsed, dried and weighed. Glycerine, sugar, magnesium sulphate, potassium sulphate, etc., will pass into solution.

(c) *Benzol, or Ethereal Extract.*—The silk, thoroughly washed and dried, is extracted with ether and weighed. The extract, being evaporated and examined, will reveal the cause of rancidity in the silk from the use of bad oils and soaps.

(d) *Action of Hydrochloric Acid.*—The sample of silk is treated for $\frac{1}{4}$ hour at 30° to 40° C. with dilute (1 : 2) hydrochloric acid. If ferric tannate has been used for loading, the reddish-yellow silk will be decolorised, and the liquid will turn a dirty brown colour, which does not become violet on addition of lime water. Should the reddish solution turn violet in presence of this reagent, logwood is indicated, and if the fibre is dark green and the liquid yellow and unchanged by lime water, then Berlin blue may be assumed. When the fibre is green and the liquid red, changing to violet when lime water is added, this indicates logwood black dyed on a ground of Berlin blue.

The iron, chrome and alumina mordants must be tested for in the liquid.

(e) *Action of Alkalis.*—After the silk has been treated as above it is boiled in an alkali solution, to dissolve the tannin, which may be detected by precipitation with iron salts.

(f) *Determination of Ash.*—A weighed sample is incinerated in a platinum crucible and calcined. If the weight is more than 1 per cent. of the original the silk has been loaded and the ash should then be further examined.

(*g*) *Determination of the Dye Stuff.*—The dye stuffs most in vogue are detected by the hydrochloric acid treatment. The further consideration of this point will be resumed later.

III. DETERMINING THE YARN NUMBER.

In comparing yarns together their thickness is employed as a means of classification, one being "coarse," another "medium," a third "fine" and so on. This method of description is, however, inexact, and it is preferable to make a comparison with a collection of standard "numbers" of yarn, by the aid of which it is easy, after a little practice, to quickly identify the number of the yarn under examination. More accurate comparison is then made by twisting the sample with the standard sample, whereby it becomes evident to a moderately good eye whether the two are of equal thickness or no.

The determination of the fineness of a yarn in figures is based upon either—

(1) *A Definite Length of Yarn.*—The various weights of this standard length are designated yarn numbers. This method of procedure is followed, as described below, in the case of silk, where the unit adopted is 9600 *aunes* (= 11,400 metres, or 12,467·4 yards), which length being weighed gives the number of the yarn.

(2) *A Constant Weight.*—In this case the various lengths required to make up this weight form the yarn numbers. This is the method pursued for all other textile fibres, *e.g.*, cotton, wool, chappe silk, etc. Of course the numbers will vary according to the system of weights adopted; for example, if the metric system is used we have metric yarn numbers, or if English weights, the English system of numbers.

For numbering purposes a certain length of yarn is wound on a reel of definite circumference, from which it is removed as hanks. These are divided into "lays" (or "leas") by means of a tie thread, each "lay" consisting of a fixed number of threads, *i.e.*, turns on the reel. Each thread is the same length as the circumference of the reel, and, when multiplied by the number of turns in the lay and the result by the number of lays in the hank, gives the exact length of yarn in the latter.

For a number of years attempts have been made to introduce a uniform system of numbering yarns but hitherto without success. In the metric system of numeration the number of the yarn denotes the number of metres that go to a gram, or kilometres to a kilogram, the length of the thread being fixed as 1000 metres with decimal sub-divisions.

The following variations above and below the exact standard representing the number of the yarn are allowed:—

1. Cotton yarns Nos. 1 to 10 English	2·5 per cent.
Waste yarn, including so-called "imitation" yarns, up to No. 6	4·0 ,,
Cotton yarns Nos. 11 to 20	2·0 ,,
,, ,, ,, 21 to 40	2·5 ,,
,, ,, above No. 40	3·0 ,,
2. Worsted yarn	1·5 ,,
3. Carded yarn	2·5 ,,
Shoddy from wool	4·0 ,,
4. Mixed wool and cotton yarn	2·5 ,,
,, wool and silk	1·5 ,,
5. Linen yarn	2·5 ,,
6. Jute yarn	3·0 ,,

In determining the number of bleached linen yarn, the loss in bleaching is fixed at 20 per cent. for $\frac{4}{4}$, 18 per cent. for $\frac{3}{4}$, and 15 per cent. for $\frac{1}{2}$ bleaching.

1. COTTON.

(*a*) *Metric Numbers.*—In this system the number of a cotton yarn indicates the number of times 1000 metres (1093·63 yds.) or $^{m}/_{m}$ required to make up a standard weight

DETERMINING THE YARN NUMBER.

of 500 grams (1·102 lbs.). For example, No. 20 cotton yarn is one, 20,000 metres (21,872·66 yds.) of which weigh 500 grams, or ½ kilo.

The following table gives the weight, in grams, of 1000 metres of the various numbers :—

No.	Weight in Grams.	No.	Weight in Grams.	No.	Weight in Grams.	No.	Weight in Grams.	No.	Weight in Grams.
1	500	15	33·333	29	17·241	43	11·628	85	5·822
2	250	16	31·250	30	16·667	44	11·364	90	5·556
3	166·667	17	29·412	31	16·129	45	11·111	95	5·263
4	125	18	27·778	32	15·625	46	10·869	100	5
5	100	19	26·316	33	15·152	47	10·638	110	4·545
6	83·333	20	25	34	14·706	48	10·417	120	4·167
7	71·429	21	23·809	35	14·286	49	10·204	130	3·846
8	62·500	22	22·727	36	13·889	50	10	140	3·571
9	55·556	23	21·739	37	13·514	55	9·091	150	3·333
10	50	24	20·833	38	13·158	60	8·333	160	3·125
11	45·455	25	20	39	12·821	65	7·692	170	2·941
12	41·667	26	19·231	40	12·500	70	7·143	180	2·778
13	38·462	27	18·519	41	12·195	75	6·667	190	2·631
14	35·714	28	17·857	42	11·905	80	6·250	200	2·500

The product of the number and weight is always the same, *i.e.*, 500, and the weight of the consecutive numbers diminishes considerably at first, but afterwards in descending ratio, and almost inappreciably in the finest numbers. The following general rules are therefore applicable to the determination of the weight and length of the threads :—

(*a*) The number of the yarn is obtained by dividing the length by twice the weight, *e.g.* :—

What is the number of a thread 60 metres of which weigh 5 grams ?—Answer : 6.

$$\text{Calculation : } \frac{60}{2 \times 5} = \frac{60}{10} = 6.$$

(β) Given the length and number of a thread, the weight is calculated by dividing the former by twice the latter, *e.g.* :—

What is the weight of 80 metres of No. 4 yarn ?—Answer : 10 grams.

$$\text{Calculation : } \frac{80}{2 \times 4} = 10.$$

(γ) The length of a thread is found by doubling the product of the number and weight, *e.g.* :—

How long is a No. 12 thread weighing 5 grams?—Answer: 120 metres.

Calculation : $2 \times 12 \times 5 = 120$ metres.

(*b*) *English Numbers.*—This system is in use not only in England but, almost without exception, also throughout the whole of Germany and Switzerland. By an English yarn number is understood the number of hanks that go to an English lb. (453 grams).

Circumference of reel = $1\frac{1}{2}$ yds. (1 yd. = 0·914 metre).

1 hank of 7 lays, of 80 turns, of $1\frac{1}{2}$ yds. each = 840 yds. (768 *m.*).

An English yarn No. 20 is therefore one of which 20×840, or 16,800 yds. (15,360 *m.*) weigh 1 lb.

TABLE OF THE WEIGHT OF 1000 METRES (1096 YARDS) OF COTTON YARN IN VARIOUS (ENGLISH) NUMBERS.[1]

No.	Weight in Grams.	No.	Weight in Grams.	No.	Weight in Grams.	No.	Weight in Grams.
4	1476	12	492	20	295	36	164
6	984	14	422	24	246	40	147
8	738	16	369	28	211	44	134
10	590	18	328	32	184	50	118

[1] Table of length in yards per lb., and weight per 1000 yards in ounces of English yarn numbers.

No.	Yards per Lb.	Weight per 1000 Yds.	No.	Yards per Lb.	Weight per 1000 Yds.	No.	Yards per Lb.	Weight per 1000 Yds.
4	3360	4·76 oz.	16	13,440	1·19 oz.	36	30,240	0·517 oz.
6	5040	3·18 ,,	18	15,120	1·065 ,,	40	33,600	0·476 ,,
8	6720	2·38 ,,	20	16,800	0·952 ,,	44	36,960	0·433 ,,
10	8400	1·90 ,,	24	20,160	0·795 ,,	50	42,000	0·380 ,,
12	10,080	1·59 ,,	28	23,520	0·695 ,,			
14	11,760	1·39 ,,	32	26,880	0·595 ,,			

DETERMINING THE YARN NUMBER. 61

The finest number of cotton yarn is No. 240, higher numbers being rarely met with in commerce. The utmost spinning capacity is about No. 300, of which $300 \times 840 = 252,000$ yards (230,400 m.) go to the lb.

Beyond 20 the even numbers only are in use—24, 26, 28, 30, and so on; in the finest yarns the numbers rise by 5 and in those above 100 by 10; Nos. 6 and 8 are the coarsest yarns. For lamp wicks Nos. ½ to 2; for tallow candles, mule yarns Nos. 8 to 12; for wax and stearin candles Nos. 20 to 40, and for woven hollow lamp wicks Nos. 12 to 30 are used. Yarns for hosiery knitting range from 6 to 36, and mule yarns 80 to 90 are also employed.

The following particulars will facilitate comparison of the English and metric systems of numbering:—

The French number $= 0.847 \times$ English No.; the International No. $= 0.423 \times$ English No.; the English No. $= 1.18 \times$ French No.; the International No. $= 0.5 \times$ French No.

It therefore follows that when a French number is compared with the English number of the same numerical value the former is finer than the latter (see following tables).

(a) COMPARATIVE TABLE OF ENGLISH, FRENCH AND INTERNATIONAL NUMBERS FOR COTTON YARNS.

Engl. No.	French No.	Intl. No.	Engl. No.	French No.	Intl. No.	Engl. No.	French No.	Intl. No.
1 =	0·85	= 0·42	26 =	22·02	= 11·01	60 =	50·82	= 25·41
2 =	1·69	= 0·85	28 =	23·72	= 11·86	62 =	52·51	= 26·25
3 =	2·54	= 1·27	30 =	25·41	= 12·7	64 =	54·21	= 27·1
4 =	3·39	= 1·69	32 =	27·10	= 13·55	66 =	55·90	= 27·95
5 =	4·24	= 2·12	34 =	28·8	= 14·4	68 =	57·60	= 28·8
6 =	5·08	= 2·54	36 =	30·49	= 15·25	70 =	59·29	= 29·65
7 =	5·93	= 2·96	38 =	32·19	= 16·1	75 =	63·53	= 31·76
8 =	6·78	= 3·39	40 =	33·88	= 16·94	80 =	67·76	= 33·88
9 =	7·62	= 3·81	42 =	35·57	= 17·79	85 =	72	= 36·
10 =	8·47	= 4·23	44 =	37·27	= 18·63	90 =	76·23	= 38·11
12 =	10·16	= 5·08	46 =	38·96	= 19·48	95 =	80·47	= 40·23
14 =	11·86	= 5·93	48 =	40·66	= 20·33	100 =	84·7	= 42·35
16 =	13·55	= 6·77	50 =	42·35	= 21·18	110 =	93·14	= 46·57
18 =	15·25	= 7·62	52 =	44·04	= 22·02	120 =	101·64	= 50·82
20 =	16·94	= 8·47	54 =	45·74	= 22·87	130 =	110·11	= 55·5
22 =	18·63	= 9·31	56 =	47·43	= 23·76	140 =	118·58	= 59·29
24 =	20·33	= 10·16	58 =	49·13	= 24·56			

(b) COMPARATIVE TABLE OF FRENCH AND ENGLISH YARN NUMBERS.

French.	Engl.	French.	Engl.	French.	Engl.	French.	Engl.
1	1·18	11	12·10	21	24·8	32	37·8
2	2·23	12	14·2	22	26·0	34	40·1
3	3·54	13	15·3	23	27·2	36	42·5
4	4·72	14	16·5	24	28·3	38	44·8
5	5·90	15	17·7	25	29·5	40	47·2
6	7·8	16	18·9	26	30·7	45	52·1
7	8·26	17	20·1	27	31·8	50	59·0
8	9·44	18	21·2	28	33·0	55	64·9
9	10·62	19	22·4	29	34·2	60	70·8
10	11·80	20	23·6	30	35·4		

Doubled yarns are counted in the same way as singles, but the number of threads is given as well, *e.g.*, for two-ply $^{40}/_2$, three-ply $^{40}/_3$, etc. In the process of doubling the yarns lose from 2½ to 6 per cent. of their length, according to the number of threads, and become correspondingly thicker. Yarns of more than two-ply are also known as sewing twist (cordonnet) and cord. A dyed, singed and finished twist is also termed lustred yarn (eisengarn).

2. LINEN YARNS.

These are divided into hand-spun and machine-spun yarns. The length and sub-division of the hanks of the former class are arranged according to greatly differing systems in different countries, but the latter are for the most part reckoned in commerce according to the Anglo-Irish system.

Circumference of reel = 2½ yds.

1 spindle, of 2 hasps, of 2 hanks, of 12 cuts (or leas), of 120 threads, of 2½ yds. each = 14,400 yds. (13,167 *m.*).

English System: 1 threescore, of 12 bundles, of 5 hasps, of 4 hanks, of 10 leas, of 120 threads, of 2½ yds. = 720,000 yds. (656,640 *m.*). This system is also in use in the linen districts of Westphalia, Silesia and Saxony.

DETERMINING THE YARN NUMBER.

The yarn number is expressed by the figure indicating the number of leas that go to an English lb.; so that, since the length of a lea is exactly 300 yards, the length of yarn going to a lb. is found by multiplying the yarn number by 300.

The numbers of linen yarns differ in comparison with those of cotton yarns because the hank of cotton measures 840 yds. whilst the length of linen cuts is 300 yds. To obtain the cotton yarn number corresponding to a linen yarn number the latter is divided by 2·8, *i.e.*, linen yarn No. 28 is the equivalent of cotton yarn No. 10. These numbers, however, represent yarns of very different appearance, the linen looking the finer.

Linen yarns are characterised as dry or wet-spun according to the method of spinning pursued. The former are possessed of a greater degree of firmness, whilst higher numbers can be obtained by wet spinning, both kinds being easily recognised by their appearance. The tow yarns, prepared from the waste in flax spinning, are very easily distinguished from linen yarn, by the numerous knots, due to contained particles of shives, which they exhibit and from which linen yarns are free.

In Germany flax is dry-spun to Nos. 10 to 30 yarns, and wet-spun up to No. 80; in Belgium and Scotland up to No. 200. Tow yarns are dry-spun from Nos. 6 to 20, and wet-spun up to No. 35, these latter yarns being used as warp for low-class fabrics, and when loosely spun and bleached, as wefts for ½-linens.

Hand-spun yarn differs from machine-spun, in handling more supple and smoother, more elastic, uneven and less rounded in form and by not rolling up, whilst machine-spun yarn feels stiffer and rougher, is of regular thickness and perfectly round.

Twist or "sewing" is the name given to yarn prepared

by twisting together two or more threads. English and Scotch twists are particularly firm and of fine appearance. The chief kinds are two and three-ply from Nos. 30 to 300 yarns; lacing twist, two and three-ply from Nos. 50 to 200 yarns; cord, three and four-ply twist from Nos. 25 to 80 yarns.

Of the other systems of numbering linen yarns in use, may be mentioned:—

The Austrian Reel: 1 schock, of 12 bundles, of 20 hanks (strähn), of 30 leas (gebind), of 40 hasps (stück), of 3 Vienna ells (ellen) = 864,000 Vienna ells. A hank, therefore, contains 3600 Vienna ells (1 Vienna ell = 0·77921 m. or 30·67749 inches). The yarn number indicates the number of hanks per 10 English lbs. (8·1 Viennese pounds).

The French Reel (partly used in Belgium): 1 shock, of 12 bundles, of 50,000 metres = 500,000 metres (546,816 yds.). Circumference of reel, 2½ metres (2·734 yds.). The yarn number gives the number of times 1000 metres (1096 yds.) going to ½ kilo. (1·102 lbs.).

3. JUTE YARNS.

Three hundred yards (274·3 m.) are wound to a lea, the number of which in 1 lb. forms the yarn number. This English system of flax numbers is used in commerce, but in the factories where spinning and weaving are carried on together the so-called Scotch system is in vogue, based on a constant length (spindle) of 14,400 yds. (13,161 m.), the number of lbs. weighed by this unit being the yarn number.

(*a*) *English Numbers.*—In England and Germany the bundle of 60,000 yds. (54,863 m.) is taken as basis. The reel circumference (thread) is 2½ yds.; 15 to 120 threads form a lea; 5 leas = 1 hank; 20 hanks = 1 reel; 16 to 20 reels = 1 bundle.

The yarns:—

No. ¼: 1 lea = 15 threads, 1 hank = 187·5 yds., 1 reel = 3750 yds.

Nos. ½-¾: 1 lea = 30 threads, 1 hank = 375 yds., 1 reel = 7500 yds.

Nos. 1-1¼: 1 lea = 60 threads, 1 hank = 750 yds., 1 reel = 15,000 yds.

Nos. 1½-12: 1 lea = 120 threads, 1 hank = 1500 yds., 1 reel = 30,000 yds.

So that—

No. ¼ contains in the bundle 16 reels, 320 hanks, 1600 leas, 24,000 threads = 60,000 yds.

Nos. ½-¾ contain per bundle 8 reels, 160 hanks, 800 leas, 24,000 threads = 60,000 yds.

Nos. 1-1¼—Per bundle, 4 reels, 80 hanks, 400 leas, 24,000 threads = 60,000 yds.

Nos. 1½-12—Per bundle, 2 reels, 40 hanks, 200 leas, 24,000 threads = 60,000 yds.

(b) *Scotch Numbering.*—1 spindle = 8 hanks = 48 leas = 5760 threads = 14,400 yds., or 1 spindle of 8 hanks, of 6 leas, of 120 threads, of 2½ yds. = 14,400 yds.

As in the case of linens, a distinction is drawn between jute line and jute tow yarns, the former being spun in Nos. 12 to 24, the coarser numbers from No. ¼ onwards in tow yarns only.

In Holland the fineness of jute yarns is given by a number denoting the number of hectograms (1 h.g. = 0·22 lb.), weight per length of 150 metres (164·4 yds.).

4. RAMIE, NETTLE FIBRE.

These yarns are numbered like chappe silk, the number denoting the number of times 1000 metres of the yarn required to weigh 1 kilo. In fineness nettle yarn No. 18 equals linen yarn No. 30 and cotton yarn No. 11 (water), and is therefore heavier than the corresponding cotton, but lighter than the linen thread of the same number. This fibre is spun up to No. 250.

5. WOOL.

Woollen yarns are divided into single and doubled yarns; and, further, according to the kind of wool and method of spinning into carded and combed (worsted) yarns. Attempts have been made to introduce a uniform metric system of numbering both kinds, but up to the present this has succeeded in the latter class only, carded yarn being still numbered according to separate standards in different countries. Thus we have English, French, Dutch, Saxon, Bohemian, etc., standards. Formerly in Austria and Germany combed yarns were reeled in the same manner as English cottons.

(a) *Metric or International System.*—1 hank = 10 leas = 730 threads = 1000 metres (10,936 yds.), or 1 hank = 10 leas = 800 threads = 1000 metres, according as the reel measures 1·37 (53¾ in.), or 1·25 (49¼ in.) metres in circumference.

The yarn number denotes the number of times 1000 metres ($^m/_m$) going to 1 kilogram, *e.g.*, No. 4 measures 40,000 metres per kilo. From this definition it follows, therefore, that, thickness for thickness, the woollen yarn number is double that of the cotton yarn number, *e.g.*, No. 40 woollen corresponds to No. 20 cotton yarn. The following rules are generally applicable:—

(a) The *weight* of 1000 *m.* of yarn is found by dividing the number into 1000.

(β) To find the number of a thread the weight is divided into 1000.

(γ) The *length* of a thread is equal to the product of the weight and yarn number.

To determine the weight of any given number, the table

given on p. 59 may be employed, except that the weight stated therein must be multiplied by 2, or else the weight of the number forming one half of the given number is taken, *e.g.*, to find the weight of 1000 *m.* of No. 20 wool the weight given in the table must be multiplied by 2, or else the number (20) must be divided by 2, *i.e.*, the weight of No. 10 cotton is the one sought.

(*b*) *English System.*—1 hank = 7 leas = 560 threads = 560 yds. (512 metres). This embodies the same basis as the cotton yarn system, *viz.*, the lb. as the unit of weight and the hank of 560 yds. as that of length. This length is termed a "conet," so that the yarn number represents the number of conets going to 1 lb. The reel is $1\frac{1}{2}$ yds. round.

(*c*) *Prussian System* :—

(*a*) 1 hasp (stück) = 4 hanks (zahlen) = 880 threads = 2200 Berlin ells = 1467 *m.* (1604·35 yds.).

Reel circumference : $2\frac{1}{2}$ Berlin ells = 1·666 *m.* ($65\frac{1}{2}$ ins.).

(β) 1 hasp = 20 leas (litzen) = 880 threads = 2150 Berlin ells = 1434 *m.* ($1568\frac{1}{4}$ yds.).

Reel circumference : 2·44 Berlin ells = 1·63 *m.* (64·17 ins.).

This is the Netherland system of reeling : in the Rhenish cloth works the following systems are preferred :—

(γ) 1 hasp = 10 leas = 1000 threads = 2000 Brabant ells = 1390 *m.* (1520·15 yds.).

Reel circumference : 2 Brabant ells = 1·39 *m.* ($54\frac{3}{4}$ ins.).

(δ) *Cockerill's reel* (also in use in Belgium) :—

1 hasp = 2240 Berlin ells = 1494 *m.* (1634 yds.).

The number indicates the number of hasps (stück) going to 1 "zollpfund" of 500 grams (1·102 lbs. Engl.), Nos. 2, 3, 4, etc., representing so many 2200 Berlin ell (or 2000 Brabant ell) lengths.

(*d*) *Saxon System* :—

(*a*) 1 hank (zahl) = 5 leas (gebind) = 400 threads (fäden) = 800 Leipzig ells = 452 *m.* (494·3 yds.).

Reel circumference : 2 Leipzig ells = 1·133 m. (44·6 ins.).

(β) 1 hank = 4 leas = 320 threads = 800 Leipzig ells = 452 m. (494·3 yds.).

Reel circumference : 2½ Leipzig ells = 1·412 m. (55¾ ins.).

(γ) 1 hank = 5 leas = 400 threads = 1200 Leipzig ells = 678 m. (741½ yds.).

Reel circumference : 3 Leipzig ells = 1·695 m. (67 ins.).

(δ) 1 hasp (stück) = 4 hanks = 12 leas = 2400 Leipzig ells = 1356 m. (1483 yds.).

(ϵ) 1 hasp = 2200 Leipzig ells = 1243 m. (1359·4 yds.).

The number indicates the number of hasps (hanks) per ½ kilo.

(*e*) *Viennese System* (current in Austria) :—

1 hank (strähn) = 20 leas (klapp) = 880 threads (fäden) = 1760 Vienna ells = 1371 m. (1499·4 yds.).

Reel circumference : 2 Vienna ells = 1·558 m. (61·16 ins.).

The number indicates the number of hanks per Viennese lb. of 560 grams (1·234 lb.).

In Bohemia 1 hank of 800 Leipzig ells is frequently taken as the standard of reeling, the number being based on the English lb. of 453 grams.

Reel circumference : 2 Leipzig ells.

(*f*) *French System* :—

(*a*) *Sedan and Neighbourhood* :—

1 hank (échevau) = 22 leas (macque) = 968 threads = 1493·6 m. (1633·45 yds.).

Reel circumference : 1·543 m. (60½ ins.).

The number indicates the number of hanks per ½ kilo.

(β) *Elbœuf* :—

1 hank = 3600 m. (3937·1 yds.).

Reel circumference : 2 m. (78·74 ins.).

The number gives the number of hanks per ½ kilo.

The following method is employed for converting the numbers of one system into those of another :—

DETERMINING THE YARN NUMBER.

From the metric number is found by multiplying—		The metric number is found by multiplying—
by 0·34	The Prussian number	by 2·93
„ 1·11	„ Saxon „	„ 0·90
„ 0·41	„ Austrian „	„ 2·45
„ 0·88	„ English „	„ 1·13
„ 0·14	„ Elbœuf „	„ 7·20
„ 0·33	„ Sedan „	„ 3·05

TABLE COMPARING THE METRIC NUMBERS FOR CARDED WOOL WITH THE SEVERAL OLDER NUMBERS.

Metric number.	English number.	Viennese.	Bohemian.	Saxon.	Berlin.	Cockerill.	Sedan.	Elbœuf.
5	4·43	2·04	4·13	2·76	1·74	1·67	1·64	0·69
6	5·31	2·45	4·96	3·31	2·09	2·01	1·97	0·83
8	7·09	3·26	6·62	4·41	2·79	2·68	2·62	1·11
10	8·86	4·08	8·27	5·51	3·49	3·35	3·28	1·39
12	10·60	4·90	9·92	6·61	4·19	4·02	3·94	1·67
15	13·3	6·12	12·4	8·27	5·23	5·02	4·92	2·08
20	17·7	8·16	16·5	11·0	6·97	6·69	6·56	2·78
25	22·2	10·2	20·7	13·8	8·71	8·36	8·20	3·47
30	26·6	12·2	24·8	16·5	10·4	10·0	8·83	4·17
40	35·4	16·3	33·1	22·1	13·9	13·4	13·1	5·56
50	44·3	20·4	41·3	27·6	17·4	16·7	16·4	6·94
60	53·1	24·5	49·6	33·1	20·9	20·1	19·7	8·33

6. SILK.

In the case of silk the fineness of the thread is determined in an opposite manner to that practised with other fibres. A sample hank of definite length is taken and weighed with weights of a special kind, the weight of the silk being known as the "titre". The higher the titre, or in other words the greater the weight of the silk, the coarser is the thread, in direct contrast to other yarns, where the fineness increases as the yarn number rises.

The sample hank measures 9600 aunes = 11,400 m. (12,467·4 yds.), the unit of weight being the *denier* = 1·26 grams (0·0464 oz.).

However, the full hank is not weighed, but only $\frac{1}{24}$th part thereof is taken as a standard cut or lea, of 400 aunes = 476 m. (520·57 yds.), and weighed with *gran*—the $\frac{1}{24}$th part of a denier or 0·053115 gram—the number of gran per cut being the same as of deniers per hank.

The single cocoon thread weighs 2 to 2½ deniers; in other silks the "denier" or fineness of the silk varies between 11 and 90, *i.e.*, a hank weighs 11 to 90 deniers. The finest organzine silk varies between 11/12 and 22/26; medium from 24/28 to 28/32; those between 30/34 and 60/70 ranking as coarse. Fine weft or trame silks range between Nos. 12/14 and 24/28; medium from 26/30 to 32/36, those between 36/40 and 70/80 being classed as coarse.

The above-named units of weight are fractions of the old Parisian lb., the old Turin lb., or the weight of the old Mailand gold mark.

1 Paris lb. of 16 oz. of 24 gran = 9216 gran, or 489·5 grams
   (1·077 lb.), so that the denier = - - - - - 1·275 grams.
1 Turin lb. of 12 oz. = 6912 gran, or 368·8 grams (0·8113 lb.) $d.$ = 1·281 ,,
1 Mailand gold mark of 8 oz. = 4608 gran, or 235 grams
   (0·517 lb.); $d.$ = - - - - - - - - 1·224 ,,

Average $d.$ = 1·26 grams.

Adopting this average value as a basis, the lengths of silk per kilo. are as follow:—

2 $d.$ = 4,528,000 $m.$ = English cotton yarn No.	2673
4 $d.$ = 2,264,000 ,, = ,,	1336
7 $d.$ = 1,294,000 ,, = ,,	764
10 $d.$ = 906,000 ,, = ,,	535
16 $d.$ = 566,000 ,, = ,,	334
24 $d.$ = 377,000 ,, = ,,	222
40 $d.$ = 226,000 ,, = ,,	134
60 $d.$ = 151,000 ,, = ,,	89
80 $d.$ = 113,000 ,, = ,,	67

Frequently the cut is made up to the round figures 480 *m.* instead of 476 *m.*, in which case the 9600 aunes = 11,520 *m.* (12,598·62 yds.).

In Italy cuts of 450 *m.* (492·14 yds.) are made and weighed with 0·05 gram; twenty-four of such cuts measure 10,800 *m.* (11,811 yds.).

In France the new titre is employed, the cut being made 500 *m.* (546·81 yds.) in length and the denier weight = 1·333 grams (0·469 oz.).

International Reel.—The International Yarn Numbering Congress, held at Vienna in 1873, decided to fix the number of silk as the number of grams weighed by a thread 10,000 *m.* (10,936·3 yds.) in length, 500 *m.* (546·8 yds.) being the unit of length and 0·005 gram the unit of weight for testing purposes.

To determine the correct titre it is necessary to wind several skeins of the standard length and weigh them, the average of the resulting figures being taken. It is always customary to give the limits of variation in weight of the individual skeins in a sample, the numbers being then written like fractions, as above.

For the conversion of the new titre into any of the older standards and *vice versâ*, all that is necessary is to multiply, or, conversely, to divide by the subjoined figures:—

Old Turin titre	0·8931
„ Mailand titre	0·9315
„ French titre	0·8964
„ Italian (also Swiss) titre	0·9000

7. CHAPPE SILK.

In Switzerland and France chappe silk is reeled in lengths of 500 *m.*:

1 skein, of 5 cuts, of 100 threads, of 1 *m.* = 500 *m.*

The number indicates the number of skeins of 500 *m.* each that go to ½ kilo.

In England chappe silk is reeled like cotton.

1 hank, of 7 cuts, of 80 threads, of 1½ yds. each = 840 yds. The number gives the number of hanks going to 1 lb.

Conversion.—By multiplying the English number by 1·69 the French, Swiss, or metric number is obtained, the converse being effected by multiplying the latter number by 0·59.

Following this necessary explanation of the various systems of numbering yarns in use, we now come to the description of the

APPARATUS FOR ASCERTAINING THE No. OF A YARN.

(*a*) *Arc Balance.*—This consists in the main of a pointer

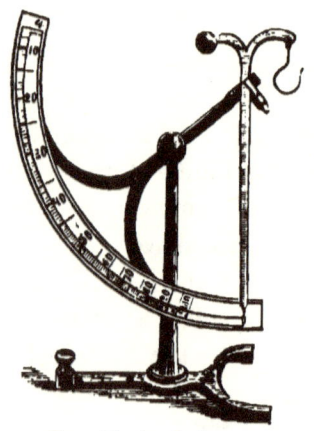

FIG. 30. Arc balance.

rotating on an axis and moving along a graduated quadrant, on which it indicates various positions of equilibrium. The yarn to be tested is affixed to the hook shown in the illustration, whereupon the number of the yarn can be read off direct from the indicator. The graduation varies of course with the system of numbering employed, so that we have cotton balances for metric and English systems, wool balances for the same systems, and so on. Moreover, various numerating scales can be attached to the same balance for the same material, an arrangement which is very convenient for the comparison of yarns.

The balance illustrated consists of an iron frame with an adjusting screw, a brass pointer, and silvered brass scale. A weight is provided for attachment to the pointer when it is desired to ascertain the number of a yarn of ten times the unit length.

(*b*) *Micrometric Yarn Balance.* — For determining the exact number of a yarn from very short lengths, *e.g.*, 4, 20,

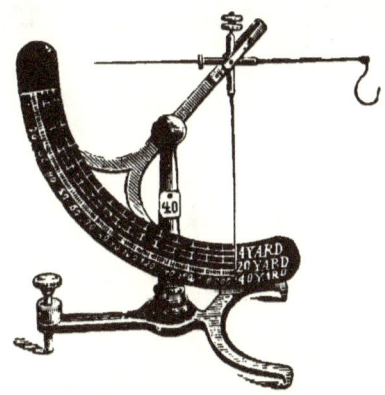

FIG. 31. Micrometric balance.

or 40 yds., or 5, 25, or 80 *m.*, without having to employ a whole reel, the following balance is used. Like the one just described, it is provided with a check weight for testing, and also with a measure ($\frac{1}{2}$ yd. or $\frac{1}{2}$ *m.*). It is very useful both in spinning and weaving works, since it enables the number of the yarn in a small piece of finished goods to be quickly and approximately determined both in warp and weft. To this end use is made of the small iron stencil plate, by means of which a small piece (100 square $^c/_m$ or $\frac{1}{16}$ sq. yd.) can be cut out of the piece to be examined, and from this sample a suitable number of threads are taken and washed to free them from dressing. These short threads (50 instead of 5 m., or 40 instead of 4 yds.) are suspended on the balance, and the yarn number is easily read off.

(c) *Horizontal Precision Balance.* — With this instrument $\frac{1}{10}$ numbers can be accurately read off quickly and without difficulty. The balance being set on a table and a screw

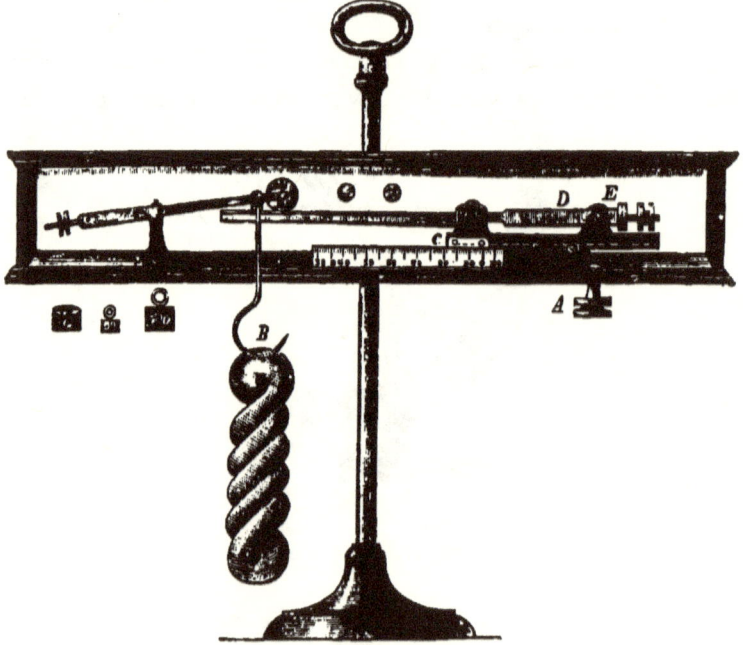

FIG. 32. Precision balance.

removed, the roller of the small balance beam is put in place resting on the larger beam, and the skein to be weighed is suspended on the hook B, the knob A being turned until the beam moves slightly, and the notch E of the beam coincides with the notch D of the index, whereupon the index shows on the scale the number of the yarn. By attaching a weight on the beam of the balance the correct number of a ten-fold length of yarn can be ascertained with the same graduation, and by interchanging weights the instrument can be used for various lengths of yarn. The entire graduation can be checked by the aid of a couple of check weights supplied with the balance.

(*d*) *Yarn Balance with Sliding Weight and Adjustable Scale Plate.*—This balance can be used for all yarns and weights; moreover, one is not tied down to any definite length of yarn,

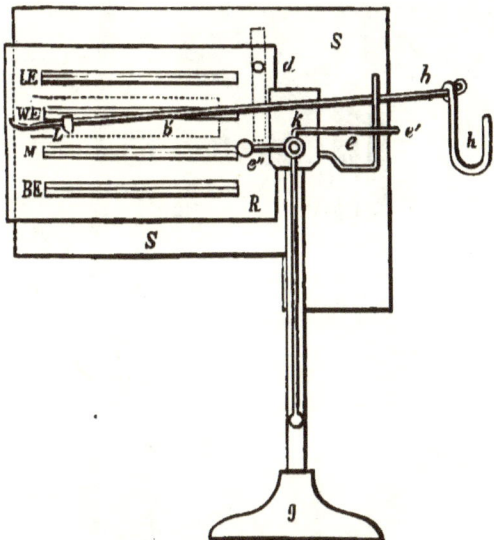

FIG. 33. Dietze's yarn balance.

but can ascertain the number from short lengths of a few inches as well as from larger ones, which in the case of fine numbers may measure several hundred yards. The yarn is hung upon a hook *h*, and the sliding weight *L* moved along the beam *bb'* until the latter takes up a position parallel to the balanced lever arm *ee'*. Each figure to which the apex of the sliding weight points on the scale must be divided into the total length of the yarn in millimetres. For instance, with a yarn 2400 *mm*. long (2·62 yds.)—which length may be made up of a number of short threads drawn from any one sample—if the weight points to 60 on the scale, the yarn number will be $\frac{2400}{60} = 40$ (metric system).

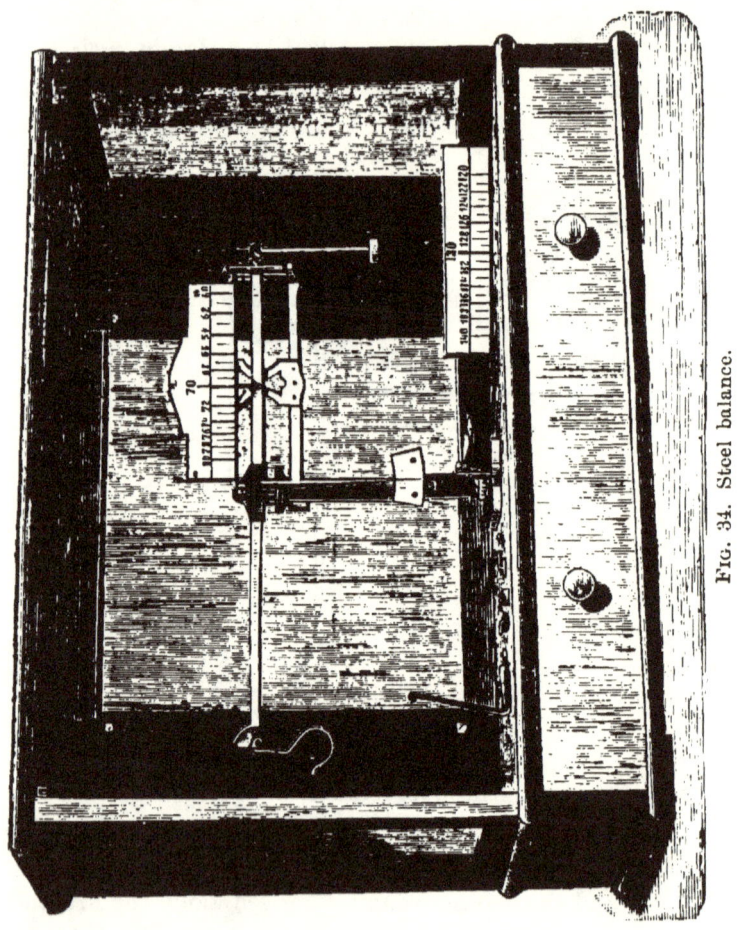

Fig. 34. Steel balance.

This method is pursued for all textile stuffs, with the exception of silk, for which the procedure is reversed, the scale number being divided by the length of the thread.

For measuring short lengths of thread a 10 *cm.* rod is

supplied, on which the yarn may be also wound in parallel layers instead of on a reel, and the skein then transferred to the balance. In this manner the number of any yarn can be very quickly ascertained.

(e) *Steel Beam Yarn Balance.*—By this balance also the number of a yarn can be very quickly determined from a definite length of thread (*e.g.*, 10, 25 or 50 *m.*), and it is particularly useful for fine spins. The arrangement can be

FIG. 35. Sampling reel.

seen from the illustration, one end of the steel beam carrying a hook for the skein of yarn, whilst the opposite arm is fitted with a sliding weight combined with a scale.

SAMPLING REELS.

In order to obtain the exact unit length of yarn for testing, recourse is had to the sampling reel. This instrument must necessarily be carefully constructed, and is therefore made entirely of metal. The arms require to be formed in such a manner that bending or alteration of any kind is out of the question. The threads are laid carefully side by side by means of a self-acting guide, so that the circumferential length is maintained unaltered throughout the winding, and, by rendering conspicuous every irregularity, enables the fineness and uniformity of the yarn to be

FIG. 36. Sampling reel.

accurately supervised. Each reel is provided with a reliable counter and bell indicator to ensure the correct number of turns being wound.

In order that the reeled yarn may be easily taken off, one of the arms is made to slide upon itself, so that, when pushed in, the yarn hangs loose; when the yarn is removed the arm is slid back to its original position and fixed there by a wedge.

Classification of sample reels for silk :—

For legal $d.$ (deniers): 400 turns × 112·5 cm. circumference = 450 m.
For old Mailand ⎫
  „ „ Turin   ⎬ $d.$: 400 „ × 119 cm.   „   = 476 m.
  „ „ Lyons ⎭
  „ new „
For international $d.$: 400 „ × 125 cm.   „   = 500 m.

Classification for Floret (chappe) silk :—

400 turns × 125 cm. circumference = 500 m. (basis 1000 m. per kilo.).

Cotton Yarn :—
English System.—Basis 840 yds. = 1 lb. 80 turns × 7 spindles × 1½ yds. = 840 yds.
French System.—Basis 100 m. = ½ kilo.
International System.—100 m. = 1 kilo.
100 turns × 7 spindles × 1·4286 m. circumference ⎫
or 100 „ × 10 „ × 1   m.   „   ⎬ = 100 m.

Worsted and Woollen Yarn :—
English System.—Basis 560 yds. = 1 lb. 80 turns × 7 spindles × 1 yd. circumference = 560 yds.
Old and New French Systems.—Basis 714 m. = ½ kilo. 100 turns × 7 spindles × 102 cm. circumference = 714 m.
International System.—Basis 1000 m. = 1 kilo.
100 turns × 7 spindles × 1·4286 m. = 1000 m.
100 „ × 10   „   × 1   m. = 1000 m.

Linen Yarn :—
Basis 300 yds. = 1 lb.
120 turns × 1 spindle × 2·5 yds. circumference = 300 yds.

IV. TESTING THE LENGTH OF YARNS.

Of no small importance in the purchase, sale and working up of yarns is the testing of the length of yarn in hanks as well as in cops and bobbins. To this end the counting reel, shown in the annexed figure, is employed. It consists of a train of wheels fitted in a box, in the cover of which are situated a pair of dial plates, one for displaying (in yards or metres) the units and the other the hundreds of the length reeled. The indicator can, by undoing a screw, be set back to zero, or at 100 or 1000, so that reversing the reel is unnecessary.

For weaving purposes, however, full length can only be reckoned upon in but few yarns, since many are reduced in length by spooling, dyeing, bleaching, etc. In this reduction

FIG. 37. Counting reel.

the dye, material and twist play a considerable part; fine and loosely spun yarns suffer more loss than stronger and better twisted ones; dark dyes, such as brown and blue, cause greater loss than paler colours. When the material is of defective composition and badly spun the reduction in length is considerable. If spun in the raw state and then dyed, the yarn shrinks in length more than if dyed before spinning; on the other hand, yarn dyed in cops and used for weft loses less than if unwound from the cop, dyed in hanks and then re-spooled for weaving.

V. EXAMINATION OF THE EXTERNAL APPEARANCE OF YARN.

For various purposes the different yarns are subjected to an examination with regard to their external appearance. Thus a combed yarn (worsted) is required to be smooth and sleek, whilst the converse qualities are exacted for carded yarn, *viz.*, a woolly surface, showing the curl of the fibre. In the case of cottons it often has to be decided whether the yarn has been previously singed or sized, and with linen yarns whether line or tow yarn is present. The latter is readily distinguished by the unequal knots (from shive residue) apparent in the thread, which should not occur in line flax. It has also to be determined whether the flax yarn has been wet or dry spun; the former being recognised by its glossy appearance, whereas dry spun yarn—usually produced in low numbers only—is dull and lustreless. In silks, raw, boiled, chappe and waste silk (bourette) yarns have to be differentiated.

The yarn is moreover examined for its fineness and regularity, for uniformity of thickness or alternating knots and weak places; for which purpose recourse is had to the instrument shown in Fig. 38, the

YARN TESTER.

This consists of an iron frame carrying a guiding screw-spindle and thread guide, the latter being moved uniformly in a horizontal direction by means of a crank on the spindle. The bobbin filled with yarn is placed on the latter, and the thread passed through the upper and lower eyelets of the guide and fastened on to the edge of a board covered with velvet, whereupon, the crank being turned, the spindle is moved horizontally, whilst at the same time the small wheel, on the axis of which the velvet-covered board is mounted, is revolved by the connecting belt at an equal rate of speed, so that the yarn is not only wound in parallel, but also in equi-

FIG. 38. Yarn tester.

FIG. 39. Yarn tester.

distant layers on the board. The dark velvet cover of the latter shows up conspicuously any inequalities in the yarn

The board may be removed by undoing a couple of screws in the metal holders, and replaced by another, so that several samples can be compared.

The yarn tester shown in Fig. 39 is arranged for two boards side by side, to take the yarn from two bobbins simultaneously.

In this test the following points have to be noted:—

In woollen yarns of equal number the thickness will vary in different samples on account of differences in the twist, a warp being finer in appearance than the same number and colour of loosely twisted weft; moreover, a dark yarn will seem to be from a half to a whole number finer than one of white wool. Furthermore, if two yarns of the same colour, but made from dissimilar wool, be held in juxtaposition, a difference in the thickness may be apparent even though they be of the same number and twist.

VI. DETERMINING THE TWIST (TORS, DRALL, DRAHT) OF YARN AND TWIST.

The smoothness of the fibre, *i.e.*, the absence of protruding hairs, is, so far as the nature of the fibre permits, influenced by the twist, increasing concurrently therewith. The "degree of twist" is indicated by the number of spiral turns imparted to the fibre within a given length.

The extent of the twist depends on:—

1. *The fineness of the yarn:* the finer the thread the greater number of twists must it receive, *i.e.*, the number of twists is in inverse ratio to the thickness of the yarn.

2. *The length of the fibres:* the longer they are the less will be the number of turns required.

3. *The object of the yarn:* that for warps is twisted more tightly than if intended for wefts, because it is subject to greater tension and abrasion in the loom, whilst weft yarn

requires to be soft and pliant in order to fill the fabric and give it the necessary closeness. Yarns for doubling are given a slighter twist than those for weaving, and yarns for cloth are made with loose twist so that they may form a felt in the fulling process: and so on.

(a) Cotton Yarns.

From the subjoined particulars it follows that the number of twists per inch can be found by multiplying the square root of the yarn number by 3·8 in the case of long staple cotton, or by 4 when short staple cotton is in question, 10 per cent. being deducted in the case of weft yarns.

In the metric system of numbering, the number of twists per 100 millimetres of the yarn is found, approximately, by multiplying the square root of the yarn number by 14 for warps, 12 for wefts, 16 to 18 for specially hard twisted water-twist yarns, or 14 to 16 for power-loom warps.

The following table gives the number of twists per inch for all cotton yarn numbers.:—

No.	Water Yarn.	Warp.	Weft.	Knit-ting.	Hosiery Yarn.	No.	Water Yarn.	Warp.	Weft.	Knit-ting.	Hosiery Yarn.
1	4·00	3·75	3·25	2·75	2·50	32	22·62	21·23	18·40	15·56	14·15
2	5·65	5·30	4·60	3·88	3·53	34	23·32	21·86	18·94	16·03	14·57
3	6·92	6·49	5·62	4·67	4·33	36	24·00	22·50	19·50	16·50	15·00
4	8·00	7·50	6·50	5·50	5·00	38	24·61	23·00	20·02	16·94	15·30
5	8·95	8·38	7·25	6·14	5·60	40	25·29	23·70	20·50	17·40	
6	9·80	9·18	7·96	6·74	6·12	42	25·92	24·30	21·06	17·82	
8	11·30	10·50	9·18	7·77	7·07	44	26·52	24·86	21·54	18·23	
10	12·63	11·84	10·27	8·80	7·90	50	28·28	26·50	23·00	19·40	
12	13·85	12·99	11·26	9·52	8·66	60	30·97	29·30	25·16	21·20	
14	14·95	14·00	12·16	10·28	9·34	70	33·44	31·35	27·17	22·71	
16	16·00	15·00	13·00	11·00	10·00	80	35·76	33·52	29·05	24·58	
18	16·97	15·90	13·78	11·66	10·60	90	37·88	35·51	30·77	26·04	
20	17·88	16·75	14·52	12·29	11·18	100	40·00	37·50	32·50	27·50	
22	18·80	17·62	15·27	12·92	11·75	120	43·80	41·06	35·58	30·11	
24	19·60	18·37	15·92	13·48	12·15	150	48·96	45·90	39·78	33·66	
26	20·40	19·12	16·57	14·02	12·75	180	53·64	50·28	43·58	36·87	
28	21·16	19·84	17·19	14·54	13·22	200	56·02	53·02	45·95	38·88	
30	21·88	20·51	17·77	15·04	13·67						

The twists can be easily determined by fastening one end of a yarn or twist thread in a vice and the other in a hand vice so that the free length of yarn measures exactly 100 mm. (3·93 in.), and then turning the hand screw, under slight tension, in the contrary direction to the twist of the thread until the latter is completely untwisted, counting the number of turns of the screw. The result will be the number of twists per 100 mm. of yarn (of course the same test can be applied to English standards by taking a yarn measuring a certain number of inches and dividing the result by that number to get the twists per inch). The more delicate forms of apparatus will be described subsequently.

Some spun yarns are "doubled," *i.e.*, two or more threads are united by twisting them together, the resulting thread being known as "twist" or "cord". The following kinds are distinguished according to their mode of preparation and uses to which they are put :—

1. Two- and three-fold twist for warp and weft.
2. Fine wefts (Nos. 80-130), two-fold for half silk.
3. Dyed and finished twist for half silk and ribbon.
4. Best twist for embroidery.
5. Multiple twist for knitting.
6. Tight and loose twists for glove making.
7. Cords for loom harness.
8. Sewing cord or twist.
9. Fancy twists, knopped, watered, glazed, and fleecy twists for modern stuffs.

Twists that are to be soft and pliant must be dry twisted, whereas those required to be dense and smooth must be twisted wet. In order that the threads composing the twist may cling closely together, they are twisted in the opposite direction to that originally employed in spinning the component yarns, *i.e.*, whilst the spun yarns are twisted in the same manner as a right-screw, the "twists" are turned like a reversed-screw.

Twist should exhibit an even turn throughout and form a sightly round and smooth thread. If an unequal tension is employed in the preparation it is easily detected, and such a twist is considered as unequally twisted.

Lustred twist is two-fold sewing twist, finished with starch and much in use on account of its strength. The turns given per 25 mm. length (1 in.) are, for:—

No. 16	-	-	17 turns.	No. 40	-	-	28 turns.
20	-	-	20 ,,	60	-	-	34 ,,
24	-	-	22 ,,	80	-	-	40 ,,
30	-	-	24 ,,	90	-	-	42 ,,

These twists are from one and a half to three times as tightly twisted as the various knitting cords.

(b) LINEN YARNS.

On account of the length of the fibre these yarns are twisted less than cottons. So, for example, Nos. 10 to 60 warp have 32 to 68 turns, whilst the corresponding wefts have 28 to 60 turns per 100 mm. (3·937 in.). The number of turns per 100 mm. is found approximately by multiplying the square root of the yarn number by 8 for line, and 8·8 for tow warps, and by 6·8 and 7·6 respectively for warps (or by one-fourth of these figures to get the number of turns per inch).

(c) WOOLLEN YARNS.

The twist given to woollen yarns depends on the quality of the wool, the fineness of the yarns and the purpose for which they are intended. A distinction is drawn between warp twist, medio twist and weft twist, the first named having to be sufficiently tight to enable the yarn to stand the tension of weaving, whilst preserving a certain elasticity. However, an excessive degree of twist may spoil a warp yarn for weaving purposes or for warping. Such over-twisting is denoted by a tendency of the yarn to curl when hanging loose in the hank. By medio twist is understood a yarn with rather more twist than weft, but employed for the same purpose. Weft yarn has a loose twist, which must, however, be sufficient to enable the yarn to stand unwinding from the bobbin.

Warp yarns are generally twisted from left to right, wefts from right to left, and although the latter are sometimes twisted in the same direction as warps, the reverse is preferable.

In the case of carded yarn the number of turns per 25 $mm.$ (1 in.) is found approximately by multiplying the square root of the yarn number by 2·58 for warps and 1·29 for wefts.

Yarn No. (metric) $mm.$	Turns per 25 $mm.$		Yarn No. (metric) $mm.$	Turns per 25 $mm.$	
	Warp.	Weft.		Warp.	Weft.
6	6	3	18	11	5·5
8	7	3·5	20	11·5	5·75
10	8	4	24	12·5	6·25
12	9	4·5	28	13·5	6·75
14	9·5	4·75	32	14·5	7·25
16	10	5	40	16	8

Woollen warps are therefore somewhat more loosely twisted than cottons, the wefts being, however, only twisted about half as much as cotton wefts. If the cloth is not to be fulled the wefts are twisted rather more tightly than otherwise. Shoddies must be more strongly twisted than yarns from natural wool.

Worsted yarns are divided—according to their employment as warp, medio twist and weft; the hardness or softness of the thread, resulting partly from the degree of twist and partly from the length and other properties of the wool—into soft worsted, middle worsted and hard worsted; also, according to the extent of cleaning to which they have been subjected, into unwashed ("in oil") and scoured worsted.

The numbering of turns per 25 $mm.$ is found approximately by multiplying the square root of the yarn number by:—

2·2 for hard twist merino warp.
1·9 ,, semi-warp (used as soft warp or as weft).
1·6 ,, soft weft.
1·2 ,, hosiery yarn from long wool.

The grades of wool employed for spinning the various yarns are denoted by letters. German spinners prepare from the subjoined wools the yarn numbers indicated opposite each.

3A (Electa wool) warp Nos. 60 to 100, weft Nos. 60 to 150
2A (Fine merino) ,, ,, 18 ,, 60, ,, ,, 18 ,, 75
A (Merino) ·
B (Purified native wool) · } ,, ,, 18 ,, 55, ,, ,, 18 ,, 60
C (Fine native wool) warp and weft Nos. 18 to 45.

In preparing doubled yarns the direction given to the twisting is, as already mentioned, the reverse of that used in spinning the component threads. If this is not done the finished cloth will have a hazy appearance that is sometimes intentionally produced for special effects by omitting the reversed twist. The number of turns is varied according to the nature of the material, the fineness of the yarn and the purpose the latter is destined to serve in the fabric.

(d) Silk.

The reeled raw silk, or grège, is not in a suitable condition for weaving and must be twisted to combine several threads into one. The first twisting (filato) is to the right, and several of these threads, being then re-twisted or doubled (torto) by a turn to the left, form "thrown" silk in contradistinction to chappe or spun silk.

According to the composition and twist of the threads silk is classified into:—

(1) *Organzine* (*Warp, or Orsoy Silk*).—From 3 to 8 cocoon threads are lightly twisted together with a right-hand twist, so that there are 60 to 80 turns per centimetre (0·3937 in.), and 2 or 3 such threads are twisted together (left twist) to form double or three-fold organzine.

(2) *Trame or Weft Silk*.—This is characterised by a much lower degree of twist; the individual threads, consisting of 3 to 12 cocoon threads, undergo no preliminary twist, and 2 or 3 of these are united by loose twisting so that the thread is softer and flatter than organzine.

(3) *Marabout Silk* is used for making crape, 2 or 3 threads being united without any preliminary twisting, then dyed without scouring, and strongly twisted. The hard twist and stiffness are characteristic of this silk.

(4) "*Soie Ondée*" is prepared by doubling a coarse and a fine thread. The material (gauze) made from this silk has a moiré ("watered") appearance.

(5) *Cordonnet*.—4 to 8 twisted threads are combined by a loose left twist, 3 of the threads thus formed being united by a right-handed twist. This silk is used for selvages, braiding, crocheting, knitting, etc.

(6) *Sewing Silk* (*Cusir*) is made from raw silk (3 to 24 cocoon) threads, 2, 4 or 6 of which are united by twisting.

(7) *Embroidery Silk*, also used for brochéing fabrics, consists of a number of simple untwisted threads united by a slight twisting.

(8) *Poil, or Single, Silk*.—A raw silk thread formed by twisting 8 to 10 cocoon threads, and employed in making gold and silver tinsel.

APPARATUS FOR DETERMINING THE TWIST OF YARN.

Twist Tester.—The instruments shown in Figs. 40 and 41 are employed for determining the number of twists in a

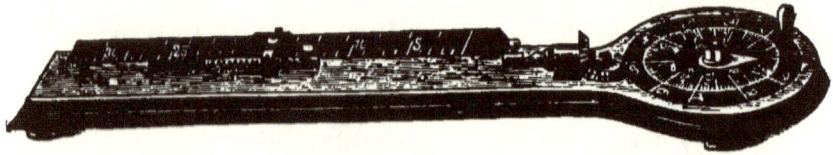

FIG. 40. Twist tester.

given length of yarn and the reduction in length it has thereby sustained. They require no further explanation.

Heal's Twist Tester.—The apparatus, made by Heal & Co. of Halifax, consists of a base plate on one side of which is erected a frame supporting two small axes. One of these axes is fitted with a crank and also carries a spur-wheel, engaging in a small spur-wheel on the upper axis, in addition to an endless screw which engages in the teeth of a

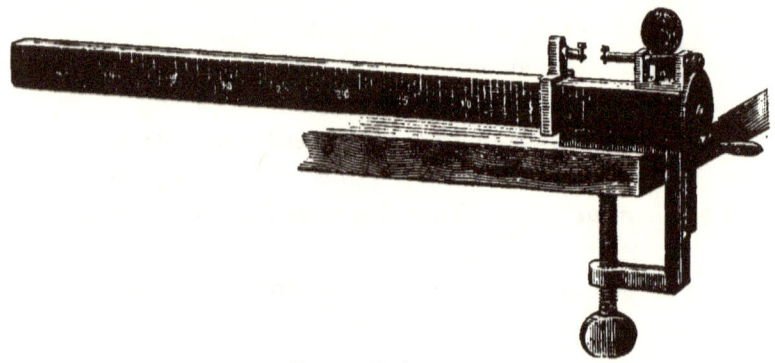

FIG. 41. Twist tester.

graduated indicator plate. The outer end of the upper axis terminates in a catch for holding one end of the yarn thread under examination, whilst the other end runs over two uprights, the further of which carries a small pulley, and the inner one a small split catch. This latter upright slides for a distance of 15 inches along a groove in the base plate, and can be fixed in any desired position by means of a thumb-screw, the adjustment being denoted by graduations on the plate.

To test the twist of a yarn the zero on the counter is adjusted to coincide with the pointer. The movable upright is placed in a position corresponding to the length of the sample of yarn, and the latter is fastened in the revolving catch, then passed over the second upright and over the pulley on the first upright, the free end being fastened to a small weight which varies in size according to the elasticity of the yarn, being heavier for stout and lighter for weak yarns. The split catch on the second upright is then

FIG. 42. Heal's twist tester.

screwed up, the lower axis set in motion by the crank and the yarn untwisted, the degree of twist being recorded by the indicator and read off direct. The indicator is arranged to register both right- and left-handed twists, and as one turn of the crank produces ten revolutions of the yarn the determination is quickly performed.

Twist Tester with Expansion Measurer and Turn Counter.—This instrument is also applicable to the determination of the twist of doubled yarn, *e.g.*, sewing cotton, sewing silk, etc. After adjusting the counter at zero the thread to be tested is fastened in the two small screw clamps, the one of which is situated to the right of the counter and the other on the left by the elasticitimeter. The latter appliance must be drawn so far out of the collar that the spiral spring of the meter is under proper tension and the indicator screw in position against the zero point on the scale. By turning the crank the doubled thread is untwisted, a needle fixed in the guide frame being inserted between the two threads.

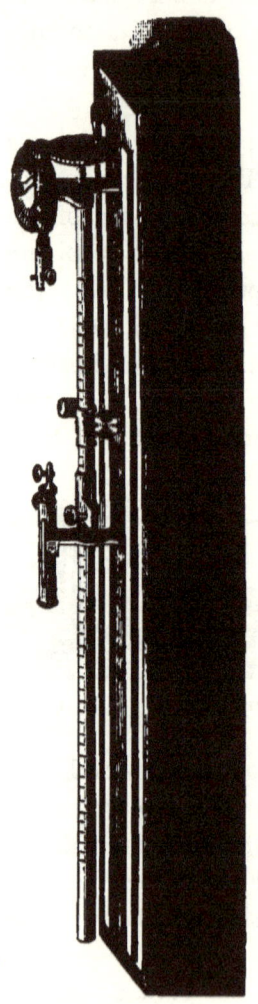

FIG. 43. Twist tester with expansion measurer and turn counter.

As the crank revolves the needle is drawn between the threads until they are completely untwisted. The counter then indicates the number of turns in the thread, while the meter shows on the left the elasticity of the thread in millimetres. The length of the test thread can be adjusted as desired, by moving the meter support along the graduated slide (50 $cm.$ in length). It is convenient to express the results on the basis of 1 metre; so, for example, if the sample thread measures 40 $cm.$, the number of turns 140, and the expansion 16 $mm.$, these results come out per metre.

$\frac{140 \times 100}{40} = 350$ turns, and $\frac{16 \times 100}{40} = 40$ $mm.$, or 4 per cent. reduction in length in twisting.

VII. DETERMINATION OF TENSILE STRENGTH AND ELASTICITY.

Tensile strength and elasticity are two important properties, both of which the spinner and weaver have to keep in mind, since their simultaneous presence is essential to a good thread. Testing by breaking with the hands is unreliable and defective, and has long been abandoned in favour of suitable mechanical testers, by means of which the spinner is enabled to check the working of his machinery and to decide what to do in case of a change in the raw material he works with. The weaver also has at disposal a reliable means of accurately comparing textile stuffs of different origin in order that he may, from the results obtained, appropriately modify the reeling and warp tension.

The tensile strength in materials of identical constitution varies inversely with the yarn numbers. So, for instance, if a No. 40 yarn has a breaking strain of 200 grams (7 oz.), then a No. 20 yarn of the same material will have a tensile strength of double that amount = 400 grams (14 oz.), and a No. 1 yarn a strength of forty times that of No. 40, *i.e.*, 8000 grams (17·6 lbs.).

These figures obtained for No. 1 yarns are known as "Quality Numbers," and give when divided by the yarn number the tensile strength in grams (or oz.), and so facilitate comparison.

The "Quality Numbers" of cotton yarns have been determined, and average—

For weak yarns	4000 grams	= 140·8 oz.
„ medium yarns	5000 „	= 176·0 „
„ strong yarns	6000 „	= 211·2 „
„ very strong yarns	7000 „	= 246·4 „
„ extra strong yarns (prima)	8000 „	= 281·6 „

To determine the quality of a yarn, from 10 to 20 breaking strain tests are made with the testing machine described below, and the mean value of the results is multiplied by the yarn number, the product being the quality.

By this means a decision can be simultaneously arrived at with respect to the uniformity of the yarn. To this end use is made of the breaking strain figures, and in this case of those falling below the mean, from amongst which a second or "sub-mean" is constructed. The difference between the sub-mean and the mean (which is most suitably expressed in percentages) gives the degree of irregularity of the yarn. For example:—

A cotton warp yarn, No. 36, gave—

Breaking strain, No. 1	=	180	grams.
,, 2	=	175	,,
,, 3	=	210	,,
,, 4	=	195	,,
,, 5	=	200	,,
,, 6	=	185	,,
,, 7	=	355	,,
,, 8	=	210	,,
,, 9	=	160	,,
,, 10	=	190	,,
Total for 10 tests	=	1860	grams.

The "mean" is therefore $\frac{1860}{10} = 186$ grams.

The tests coming out below this mean were:—

No. 1	=	180	grams.
,, 2	=	175	,,
,, 6	=	185	,,
,, 7	=	155	,,
,, 9	=	160	,,
Total for 5 tests	=	855	grams.

The "sub-mean" is therefore $\frac{855}{5} = 171$ grams, and the difference between mean and sub-mean 15 grams, which, expressed in percentages $= \frac{15 \times 100}{186} = 8$ per cent.

DETERMINATION OF TENSILE STRENGTH AND ELASTICITY. 95

The *strength* of the yarn therefore was 186 grams (6·54 oz.), and its *irregularity* 8 per cent.

By practical experience it has been determined that when the difference between the mean and sub-mean varies—

>Below 10 per cent. a yarn is considered as very uniform.
>„ 15 „ „ „ uniform.
>Above 15 „ „ „ irregular.

The strength of single cotton yarns is given in the subjoined table in grams (= 0·3527 oz.) :—

No.	Weak.	Medium.	Strong.	Very Strong.	No.	Weak.	Medium.	Strong.	Very Strong.
4	880	1000	1250	—	32	125	170	200	250
6	670	920	1080	1340	34	120	160	190	220
8	500	690	810	1000	36	110	150	180	210
10	400	550	650	800	38	105	140	170	200
12	330	460	540	660	40	100	135	160	190
14	285	390	460	570	50	—	110	130	140
16	250	340	400	500	60	—	90	110	125
18	220	300	360	440	70	—	80	90	105
20	200	280	320	400	80	—	70	80	95
22	180	250	295	360	90	—	60	70	85
24	170	230	270	330	100	—	55	65	80
26	150	210	250	310	110	—	50	60	70
28	140	200	230	290	120	—	45	55	60
30	130	180	215	260					

The strength of flax yarns is easily calculable from the subjoined formula :—

Let G = strength (grams or oz.), N = yarn number; then, for tow yarns for hand-looms

1. $G = \dfrac{19{,}000 \text{ grams}}{N}$; 2. $G = \dfrac{21{,}000 \text{ grams}}{N}$, *e.g.*, the strength of No. 16 yarn = 1187 grams (41·78 oz.); or 1312 grams (46·18 oz.).

In English four-fold sewing twist: $G = \dfrac{21{,}385 \text{ grams}}{N}$.

In string from fine-hackled flax: double twist: $G = \dfrac{31{,}316 \text{ grams}}{N}$; three-fold: $G = \dfrac{32{,}500 \text{ grams}}{N}$; $\dfrac{35{,}780 \text{ grams}}{N}$; $\dfrac{41{,}000 \text{ grams}}{N}$; the mean of six grades of string being: $G = \dfrac{35{,}000 \text{ grams}}{N}$.

By moderate twisting the strength of the thread is increased, but if twisted to excess it becomes brittle and loses its elasticity.

The *elasticity* of a yarn is expressed by the increase in length it undergoes when strained to breaking point. This is determined in a simple manner by making one end of a half-metre (19·68 in.) length of yarn fast, and attaching the other to a drum 5 *cm.* (2 ins.) in diameter, the axis of which is at a distance of half a metre from the other extremity of the yarn. By means of a crank this drum is revolved until the thread breaks, the arc through which the periphery of the drum has moved, and which is ascertained by the graduated markings on the same, indicating the elasticity of the yarn.

The elasticity of cotton yarns should be about:—

For Nos. 20 to 30	4·5 to 5	per cent.
,, 30 ,, 40	4 ,, 4·5	,,
,, 40 ,, 60	3·8 ,, 4	,,
,, 60 ,, 80	3·5 ,, 3·8	,,
,, 80 ,, 120	3·0 ,, 3·5	,,
,, 120 ,, 140	2·5 ,, 3·0	,,
,, 140 ,, 170	2·0 ,, 2·5	,,

The following particulars of the tensile strength and elasticity of wool fibres are given by Bowman:—

Kind of Wool.	Breaking Strain.		Elasticity. Per Cent. of Length.	Diameter of Fibres. Inches.
	Grms.	= Oz.		
Leicester Wool	32·63	1·15	0·284	0·00181
South Down Wool	5·59	0·20	0·268	0·00099
Australian Merino	3·25	0·114	0·335	0·00052
Saxon Merino	2·54	0·0894	0·272	0·00034
Mohair	38·09	1·34	0·299	0·00170
Alpaca	9·68	0·337	0·242	0·00053

The elasticity of silk amounts to a seventh to a fifth of the length. Raw silk is more elastic, and withstands strain better than scoured silk, since, in the operation of boiling,

DETERMINATION OF TENSILE STRENGTH AND ELASTICITY. 97

about 45 per cent. of the elasticity and 30 per cent. of the tensile strength are lost. Wet silk has more elasticity but less tensile strength than dry.

On the proposal of Reuleaux the definition of "breaking length," *i.e.*, the length of a thread, at which the thread itself is broken by its own weight, was adopted.

The accompanying table has been compiled from the results of experiments:—

	Breaking Length in Kilometres (1093·6 yds.).	Tensile Strength in Kilograms per sq. *mm.*		Breaking Length in Kilometres.	Tensile Strength in Kilograms per sq. *mm.*
Sheep's Wool	8·30	10·9	Vegetable Silk	24·5	—
Cocoanut Fibre	17·8	29·2	Cotton	25·0	37·6
Jute	20·0	28·7	Hemp	30·0	45·0
China Grass	20·0	—	Manila Hemp	31·8	—
Flax Fibres	24·0	35·2	Raw Silk	33·0	44·8

APPARATUS FOR TESTING THE BREAKING STRAIN OF YARN.

1. *Pocket Tester* (Ulmann, Zurich).—A spiral spring is enclosed in a wide, round or elongated cylindrical case (Figs. 44, 45). To the end of the spring is attached a hook, on to which the end of the yarn, under examination, is fastened, and the instrument being held or suspended by the ring at the top the yarn is pulled until it breaks. In the watch-shaped tester (Fig. 44) the black pointer will fly back to zero at the instant the yarn breaks, leaving the white pointer to indicate on the scale the tensile strength of the yarn in grams. This white pointer must, of course, be adjusted to zero before commencing the next test.

The tester shown in Fig. 46 consists of a cylindrical case enclosing a spiral spring, the box containing the instrument being convertible into a stand for same. The thread to be tested is pressed down on the table with the thumb, fastened on to the hook, and the end drawn out by the other hand until the yarn breaks. The breaking strain is expressed in weight by the position of the indicator on the scale.

2. *Breaking Strain Tester for Simple Threads* (without the elasticitimeter).—The instrument consists of a stand on

Fig. 44. Pocket instrument for testing the strength of yarn.

Fig. 45. Pocket instrument for testing the strength of yarn.

Fig. 46. Breaking strain and elasticity tester.

which is mounted an arc graduated in front, as a scale, at intervals expressing 10 grams weight, the upper edge of the arc being cut as a toothed rack. The stand also carries a lever, one arm of which terminates in a hook for holding the yarn, whilst the other, forming the pointer, is straight and fitted with a small detent. To apply the test the base screws are adjusted until the instrument is upright; the yarn is

DETERMINATION OF TENSILE STRENGTH AND ELASTICITY. 99

attached by one end to the hook, and the other end, after being passed over a small roller at the base, is pulled gradually until it breaks, the detent engaging in the teeth of the arc as the pointer moves, and detaining the latter in the position it has assumed when the breaking point is reached, thus enabling the weight indicated to be read off at convenience. The test should be repeated about ten times in order to obtain the true average breaking strain.

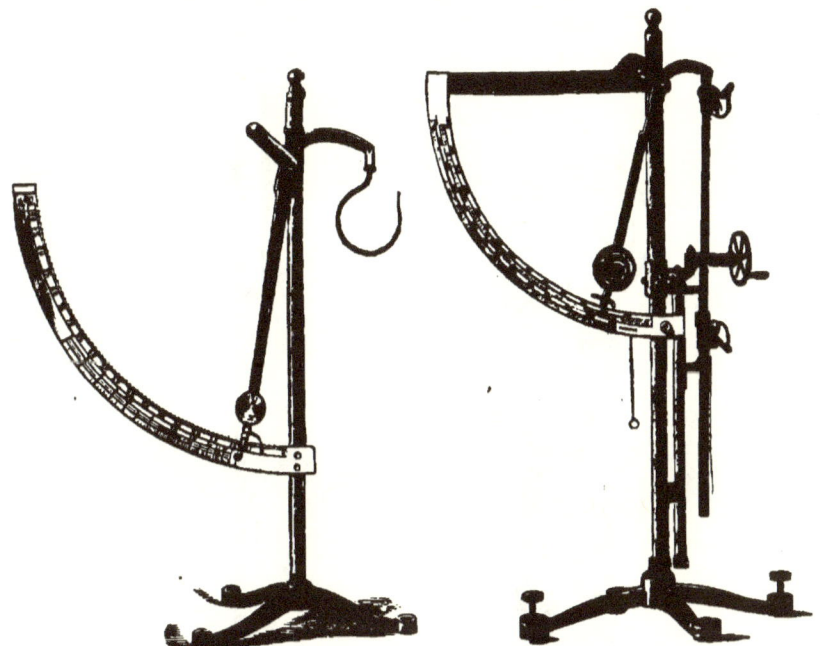

FIGS. 47 and 48. Breaking strain tester.

3. *Elasticity Tester.*—A small instrument for this purpose is shown in Fig. 49. This is screwed on to a table, and the thread to be tested is wound round the horizontal roller, one end being held fast whilst the other is pulled until the thread breaks, whereupon the position of the indicator will show the elasticity of the yarn in millimetres.

100   YARNS AND TEXTILE FABRICS.

4. *Combined Breaking Strain and Elasticity Tester* (Fig. 50).
—This instrument is fastened on to a table by means of the

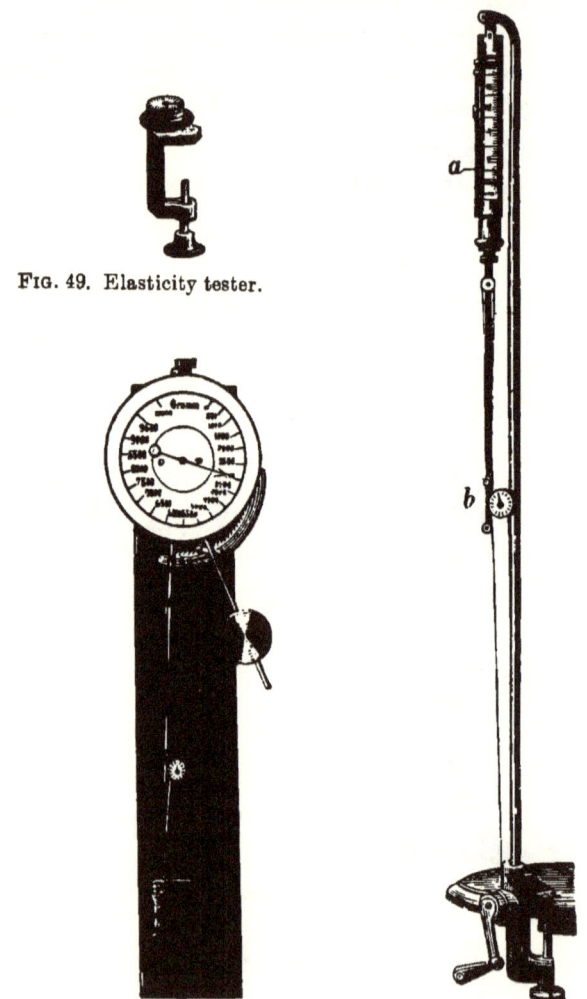

FIG. 49. Elasticity tester.

FIG. 51. Breaking strain tester.   FIG. 50. Combined breaking strain and elasticity tester.

clamp, the yarn being then twisted around the small upper roller and passed between the roller b and the two small

DETERMINATION OF TENSILE STRENGTH AND ELASTICITY. 101

rollers shown in the Fig.; the lower end passing down to the crank axis, to which it is attached by a clamping screw. The crank is turned until the thread breaks, whereupon the indicator a will show the breaking strain in grams, and the scale will record the elasticity of the yarn in millimetres. Both the indicators a and b should be set back to zero before commencing a test.

The tester shown in Fig. 51 is intended for the same purpose as the others, but can be used for testing 10 or more threads at a time. The apparatus is fixed on a board, and carries a large dial plate on which the scale is marked, a

FIG. 52. Piat and Pierrel's breaking strain tester.

weight being used, in place of a spring, for determining the breaking strain. The scale is usually graduated up to 10 kilos. (22 lbs.). If doubled yarns are tested the result must be divided by the number of component threads, in order to obtain the figures denoting the individual strength.

In the case of doubled threads, the strength is increased by the operation of doubling.

5. *Piat and Pierrel's (St. Maurice) Breaking Strain Tester.*—As can be seen from Fig. 52, the instrument consists of two parts, that on the left being the indicator of the tension

produced by the right-hand portion of the apparatus. The thread is stretched by turning the crank, whereupon the lever inclines and the pointer indicates on the graduated quadrant the tension on the thread at the moment of breaking, a detent restraining the lever in its actual position at the time; this gives the breaking strain.

The elasticity stands in direct relation to the stretch of the thread, and is measured by the distance travelled by the detent along the toothed quadrant.

Of course, the test must be performed on a number of threads. The test weight is made heavier or lighter according to the strength of the yarn, and the figures obtained for the breaking tension must in each case be multiplied by the weight employed.

6. *Schopper's Yarn Tester.*—The apparatus described below was designed in response to an invitation by the Royal Experimental Institute at Charlottenburg. The arrangement can be seen from Fig. 53, and the test is applied as follows: After the loaded lever G has been adjusted to point to zero, the upper tension clamp J is held back by the catch provided for that purpose, and the lower clamp M raised to its highest position, by turning the hand wheel, and fixed in place by means of the screw X, situated below the nose lever O. By pushing the tension rod case up as far as possible, the expansion lever is adjusted at zero, and the thread is then fixed in position, an operation more easy of performance with this vertical apparatus than with horizontal instruments. The upper clamp is then released from its catch and the loaded lever also set at liberty; and the clicket d being lowered, the apparatus is set in work by turning the hand wheel B. After two or three turns of the wheel, and without stopping, the lower clamp is released by undoing the screw X, so that when the thread breaks this may fall and so release the expansion lever. When breakage

DETERMINATION OF TENSILE STRENGTH AND ELASTICITY. 103

occurs both the levers remain in their actual position at the moment—whether the hand wheel is stopped or not—and so allow the breaking weight and the expansion to be read off with accuracy.

If, in addition to the breaking strain, the breaking length

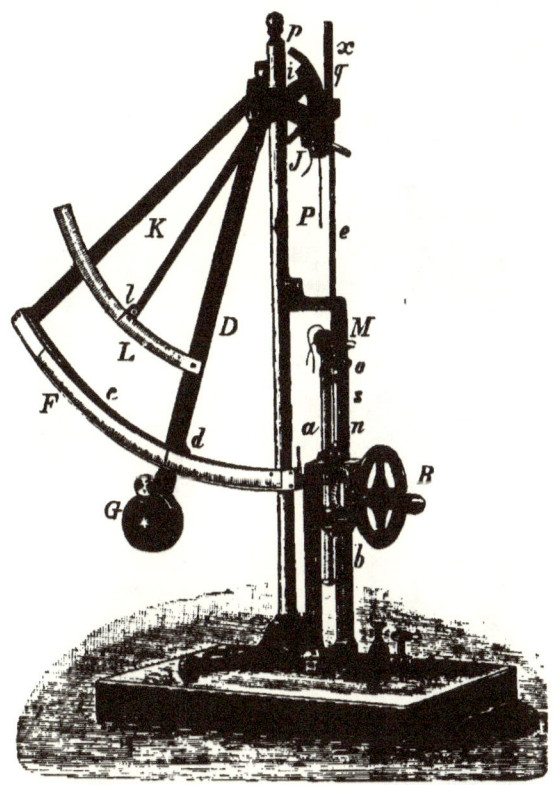

FIG. 53. Schopper's breaking strain tester.

has to be determined, both ends of the broken thread are cut off short at the clamps and weighed together. The calculation of the breaking length is performed according to the well-known formula :—

$$R \text{ (breaking length)} = \frac{L \times B}{G}$$

wherein L = length of the broken thread; G = weight of the same; and B the weight representing the breaking strain.

In setting up this instrument care must be taken that

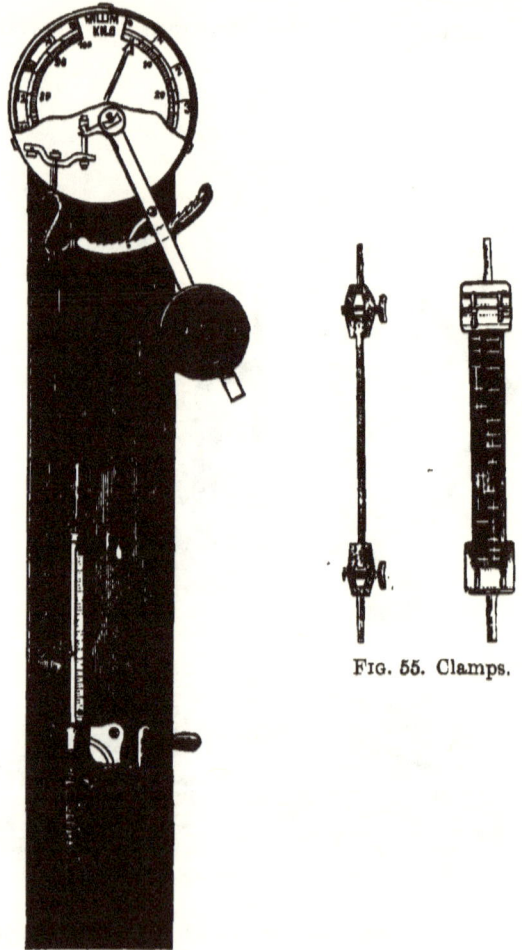

Fig. 54. Breaking strain and elasticity tester.

Fig. 55. Clamps.

the indicator of the weight lever is adjusted to zero and that the air bubble in the level at the foot of the stand is exactly in the centre.

7. *Breaking Strain Tester with Elasticity Tester* (Fig. 54).—This apparatus is also arranged to work by dead weight instead of by a spring, and is used for testing whole and half cuts of string, cord, etc., the test being very quickly performed. The skein under examination is attached to sufficiently strong hooks both above and below, and the crank with attached gearing set in motion until breakage occurs, the strain being indicated on the dial plate, where the expansion is also recorded in millimetres. To ascertain the true elasticity of the thread, the figure recorded on the scale attached to the lower hook is deducted from that shown on the dial. The instrument is mounted on a stout oaken board, and all the parts are constructed of steel and brass.

A large quantity of yarn is required for the test, but the results are characterised by greater accuracy. In attaching the yarn to the hooks the turns should lie side by side so that they may slide freely when stretched. Instead of hooks, clamps (Fig. 55) may be fitted to the machine, which is thus made available for testing cloth as well as yarn. The driving wheel may be arranged to work by hand or power.

8. *Continuous Tester for Determining the Elasticity and Strength of Yarn.*—This apparatus (invented by Holzach and made by Wenner of Zurich), which is of great service to the spinner, indicates automatically and simultaneously both the strength and elasticity of the yarn which is being passed through it. Being driven by belting, a considerable length of yarn (some 15 metres per minute) can be tested in a short time. The thread runs from the bobbin through the guide L on to the pair of conical rollers A, and being held fast by the upper pressure roller (covered with rubber or leather), passes over the pulley of the spring balance S and is drawn by the pressure roller of the pair of cylinders B (worked

by the intermediate gear *P*—speed, 120 turns) and wound on a plush roller. The guide *L* can be slid along a scale indicating the percentage of expansion sustained at any moment by the thread. The two pairs of rollers are of different diameter, but are driven at an equal number of turns. The farther the guide *L* is pushed towards the

Fig. 56. Holzach's continuous tester.

smaller end of the cone, the greater is the difference traversed by the thread over *B* as compared with *A*, the maximum of difference giving the expansion of the thread at the time of breaking. The breaking strain is indicated by the balance *S*, and the machine is thrown automatically out of gear when the thread gives way.

VIII. ESTIMATING THE PERCENTAGE OF FAT IN YARN.

This determination has to be made in the case of worsted, carded wool, and artificial wool yarns, and can be performed in a Soxhlet extractor by means of petroleum spirit, the simplest method of procedure being, however, as follows: 5 grams of yarn are steeped with 100 *c.c.* of petroleum spirit in a flask, well shaken up and left to stand for a day, whereupon 5 *c.c.* are syphoned off and the spirit removed by evaporation in a tared basin over the water bath.

The employment of petroleum spirit is preferable to using other solvents such as water-free ether or carbon bisulphide.

When the Soxhlet extractor is used, 2 to 5 grams of the material are placed in the widened tube, which is fixed on to a small flask (weighed when dry), the latter being half full (about 50 to 70 *c.c.*) of ether, and the apparatus is surmounted by an inverted condenser in which the ascending ethereal vapour is condensed and continuously falls back into the extractor. After the apparatus and its contents have been heated for some time over the water bath, all the oil and fat in the substance will have passed into solution in the ether, the flask being then removed and the ether distilled, the residual fat being dried for a half to three-quarters of an hour at a moderate temperature, and weighed when cold. The increased weight of the flask gives the amount of oil and fat present in the sample.

IX. DETERMINATION OF MOISTURE.

CONDITIONING THE YARN.

All fibres, whether of animal or vegetable origin, are endowed with the property of hygroscopicity, *i.e.*, power of absorbing water from the air, and undergo alterations in weight and volume corresponding to the amount of moisture so taken up. In damp air they expand and become heavier, and in dry air contract and lose weight, the cause of this phenomenon being due to the variable temperature of the atmosphere. According to the origin of the fibre the quantity of moisture absorbed and retained is larger or smaller. By reason of this peculiarity insurmountable difficulties would attend commercial dealings in textile fibres, especially the more expensive products of silk, were there no means of determining and taking into account the percentage of moisture they contain.

At the request of the parties interested, the maximal admissible percentages of moisture for the various fibres, and the normal amounts present therein in the air-dry condition, were experimentally ascertained some fifty years ago. These investigations led to the installation of testing institutions, the so-called "conditioning" establishments, of which there are now some thirty-two in existence for the examination of silk; *e.g.*, Lyons, Paris, Mailand, Marseilles, Florence, Turin, Vienna, Zurich, Basle, Crefeld, Elberfeld. For a number of years attempts have been made in Germany to have the condition of wool and woollen yarns officially determined under legislative control.

So far as the vegetable fibres are concerned, it is but seldom that any account is taken in commerce of their content of moisture, although, as is shown in the subjoined tables, the amount of water fixed by the various fibres is not unimportant.

TABLE OF WATER CONTENT OF DIFFERENT FIBRES, IN AN AIR-DRY CONDITION, AND AFTER EXPOSURE FOR SOME TIME IN AN ATMOSPHERE SATURATED WITH MOISTURE.

Fibre.	Water content in air-dry condition. Per cent.	Maximum percentage absorbed.	Fibre.	Water content in air-dry condition.	Maximum percentage absorbed.
Cotton	6·66	20·99	Manila	12·50	40·00
Flax (Belgian)	5·70	13·90	Jute	6·00	23·30
" "	4·20[1]	24·00[2]	Wool	8-12	30-40
Nettle	6·52	18·15	Silk	10-11	30

The International Congress for the establishment of a uniform system of numbering yarn, held at Turin, agreed upon certain admissible normal percentages of moisture corresponding to the peculiarities of the various fibres. These quantities are known as the "reprise," *i.e.*, the amount of moisture present in the fibre under ordinary atmospheric conditions, or the amount reabsorbed by the fibre on cooling in the air after having been subjected to prolonged drying in hot air.

	Increase compared with the dry weight per cent.	Percentage of moisture in total weight.
Cotton yarn, incl. "imitation" yarn	8½	7·83
Silk thread	10	9·09
Flax yarn	12	10·71
Hemp yarn	12	10·71
Tow yarn	12½	11·11
Jute yarn	13¾	12·09
Jute yarn (according to Pfuhl)	14	12·28
Carded woollen yarn	17	14·53
Worsted	18¼	15·43
Yarns from wool and silk	16	—
Yarns from wool and cotton	10	—
Shoddy yarns	13	—

[1], [2] Figures given by the Trautenau Experimental Station for the cultivation of flax.

CONDITIONING SILK.

After the bale of silk has been weighed, some 18 to 30 skeins of a total weight of about 1½ kilos. (3·3 lbs.) are taken, some from the centre and others from near the outside, and divided into three equal samples, which are accurately

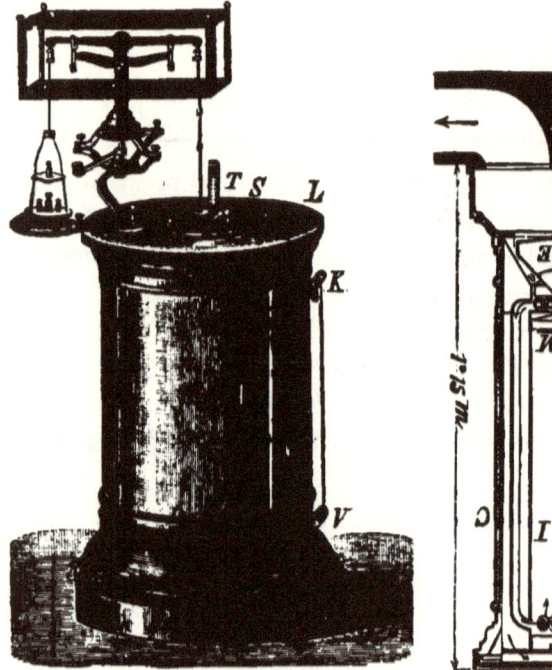

Fig. 57. Conditioning apparatus.

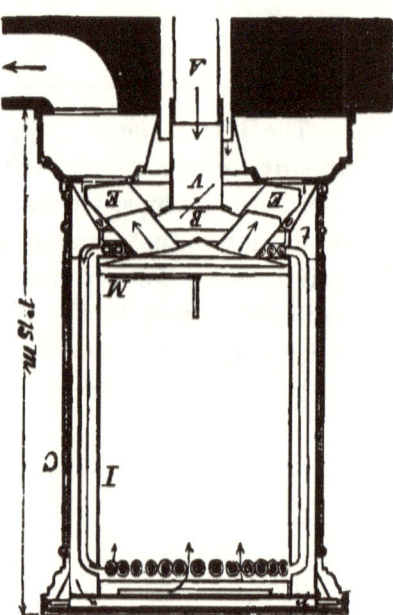

Fig. 58. Sectional view.

weighed. Two of the samples are then placed in the conditioning apparatus described and illustrated above (Figs. 57, 58), and are dried for several hours at a temperature of 110° C. The weight is taken at intervals of twenty minutes, and as soon as two weighings differ by not more than ½ per cent. the weight is calculated to that of the entire bale, with the addition of the legally permissible limit of moisture, *viz.*, 11 per cent. The third sample is used in the event of any difference resulting.

The conditioning apparatus invented by Talabot of Lyons in 1841, and improved by Persoz and Rogeat, is constructed as follows:—

A constant temperature of 110° C. is maintained in the apparatus by means of two currents of air, one merely lukewarm and the other heated to over 110° C., supplied from a heating stove in the cellar of the building. The hotter current enters through the tube A into the chamber B, where it divides and passes through thirty-two perpendicular tubes $l\ l$, and emerges thence into the actual drying chamber I, in which the silk skeins to be tested are suspended freely from hooks attached to one arm of the balance forming part of the apparatus. The cool, dry air current enters the apparatus through the annular passage surrounding the tube A, plays round the tubes $l\ l$, and enters the chamber through an opening, r, in the cover. The two currents can be restricted, increased, or shut off altogether, by suitably adjusting the valve v and the valve controlling the aperture r, without it being necessary to open the apparatus; and the production of an atmospheric mixture with a constant temperature of 110° C., indicated by the thermometer inserted in the apparatus, is thereby facilitated. This current of air surrounds the silk, absorbs its moisture, and, laden with the latter, passes out of the chamber into the chimney shaft *via* the flues $E\ E$. In this manner the silk is dried rapidly, the absolute dry weight being obtained in a half to three-quarters of an hour. The air is shut off, by closing the valves, whilst the weighings are in progress.

CONDITIONING LOOSE WOOL.

The process will be explained on the basis of a 2000 kilos. (4408 lbs.) parcel of wool sent to be conditioned. The bales are weighed directly on reception, and as accurately as possible, and samples are at once taken from various parts of the different bales. When the bales weigh 120 to 150 kilos. (264 to 330 lbs.), the samples taken from each must be not less than 1 to 1½ kilos. (2·2 to 3·3 lbs.) in the aggregate; and a conditioning test must be performed on each 400 kilos. (880 lbs.) of wool. The different samples are united, and from the whole are drawn three test samples of 500 grams (1·1 lbs.), of which two are placed in the apparatus. If, after a short time, the loss in the two samples agrees or differs by barely ½ per cent., the test is at an end, but if the difference is greater the third sample is tested, and the mean of the results is taken and expressed in percentage.

In a 2000 kilos. parcel from ten to fifteen samples are thus taken, the average percentage loss in which forms a basis of calculation for the entire parcel. If the latter consist of thirteen bales, and a 1½ kilos. (3·3 lbs.) sample be drawn from each, then we have 19½ kilos. (42·9 lbs.) of sample, which may be assumed to be representative of the bulk.

From this are taken fifteen samples of 1500 grams = 7500 grams (16½ lbs.), which, when conditioned, yield a dry weight of say = 6300 grams (13·86 lbs.), corresponding to an average loss of = 1200 grams (2·64 lbs.), or 16 per cent., so that the dry weight in 100 lbs. of the wool would be 84 lbs. of actual wool. Now, since the permissible limit of moisture in wool is 17 per cent., there must be added to the 84 per cent. of wool 14·28 per cent. as permissible moisture, which gives the normal weight = 98·28 per cent., and the inadmissible water = 1·72 per cent. Therefore, instead of 2000 kilos. (4400 lbs.), the weight of the parcel must be reckoned as

$$2000 - \frac{2000 \times 1\cdot 72}{100} = 1965\cdot 60 \text{ kilos. } 4324\cdot 32 \text{ lbs.}).$$

DETERMINATION OF MOISTURE.   113

CONDITIONING WORSTED.

The process is identical with the foregoing, but a few preliminaries have to be gone through. When the yarn is on bobbins or conets, it must, after the gross weight has been ascertained, be reeled in hanks of 1000 metres (1093·6 yds.), and the tare deducted. If the yarn contains any matters that volatilise during the drying, it is first washed and freed from fat, and then transferred to the conditioning apparatus.

APPARATUS.

The Kohl Conditioning Apparatus for wool consists of a delicate balance (constructed for a maximum load of 1 kilo.

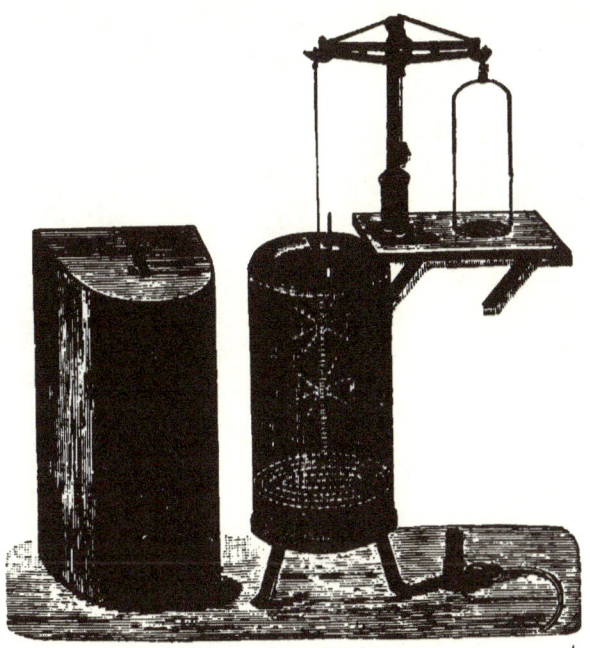

FIG. 59. Kohl's wool conditioning apparatus.

(2·2 lbs.), and capable of indicating down to 1 centigram); a strong copper drying vessel fitted with a thermometer; an

114   YARNS AND TEXTILE FABRICS.

iron case; a drying rack for cops; another for sliver; and a pair of gas burners or petroleum stoves.

The drying frame suspended from the balance is filled with a number of cops or sliver, the weight of which is accurately determined at the outset, and again after one and a half to two hours' drying, a third weighing being made

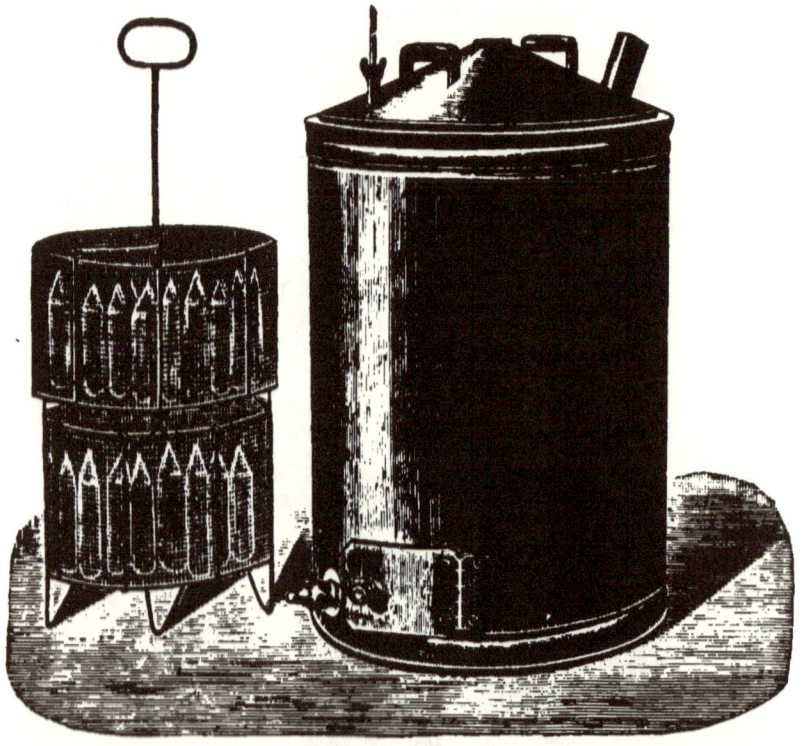

FIG. 60. Heal's conditioning oven.

after the lapse of a further half-hour. If no loss of weight occurs between the second and third weighings the final weight, with the addition of the permissible percentage of moisture, is calculated to the total amount of the wool.

Heal's Yarn Testing Oven (Fig. 86) consists of a cylindrical vessel fitted with a conical movable cover, and containing an inner vessel of slightly smaller diameter, the space between them being closed at the top but open below. At the bottom of the inner vessel there is a plate which is exposed to the flame of a Bunsen ring-burner, the heated gases from

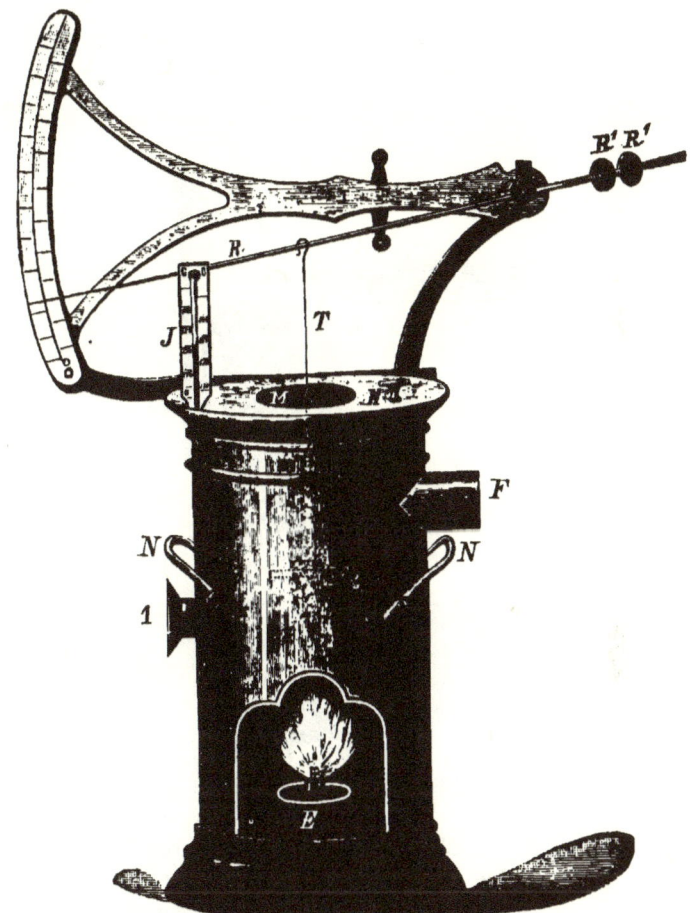

FIG. 61. Ulmann's conditioning apparatus.

which surround the inner vessel, the products of combustion escaping through the flue shown at the right-hand side of the apparatus. The cops or raw material to be tested are laid in a circular wire basket made in two parts in order to ensure an equal distribution of heat through the cops. The material is of course weighed before being placed in the basket. The temperature of the apparatus is registered by a thermometer inserted through the lid.

Conditioning Apparatus with Arc Balance (Ulmann, Zurich). —This useful apparatus is shown in the annexed figure (Fig. 61). The method of manipulation is as follows: The

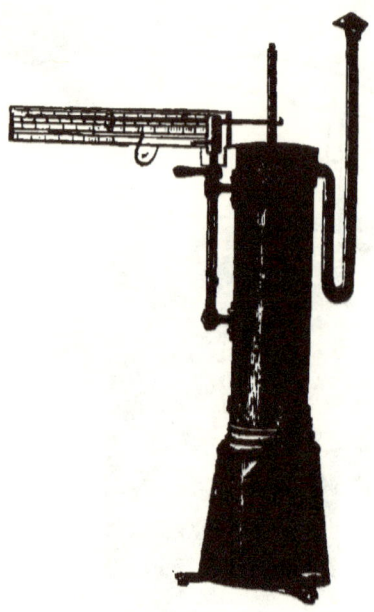

FIG. 62. Conditioning apparatus with sliding weight.

rod T supports a basket inside the apparatus, in which 20 grams (0·7 oz.) of the sample to be tested are laid, care being taken to prevent contact with the sides of the apparatus. The lid M is then closed, and the weights R^1 are moved along the threaded arm until the pointer is opposite zero on the scale. The spirit or gas lamp E being then lighted, heat is applied until the thermometer J registers

110° to 120° C., whereupon the internal valve, which is left open at the commencement of the test, is more or less closed by turning the handle H, in order to regulate the admission of hot air and maintain the temperature at the requisite degree. The moisture will, as a rule, all be expelled in about twenty to thirty minutes, the lamp being then extinguished, and the final position of the pointer, which will have risen, indicates, on the scale—which is graduated in centigrams—the amount of moisture in the sample examined. The oven, which is constructed in two sizes, for conditioning 20 and 100 grams (0·7 and 3·5 oz.) respectively, is of the following dimensions: Diameter, 23 or 25 *cm.* (9¼ or 10 in.); height, 48 or 68 *cm.* (19 or 27 in.).

The Findeisen (Chemnitz) Conditioning Apparatus (Fig. 62) consists of an oven, a copper vessel warmed by a gas or spirit lamp, and a delicate sliding-weight balance, the pan of which is placed inside the oven. When the sample is placed in the pan the apparatus is heated to 110° to 120° C. for some time, the dry weight of the substance being ascertained by counterpoising the balance by sliding the weight along the beam.

TESTING MANUFACTURED FABRICS.

"Fabric" is the name given to any material prepared by the regular interweaving of threads or thread-like bodies by the aid of mechanical appliances. This general classification is subdivided into "weavings" or woven material, composed of threads interwoven at right angles, and worked goods, wherein the threads are combined in serpentine fashion or so twisted as to form a mesh. In this latter category must be included machine lace, tulle, bobbinet, and such like materials, in addition to those produced in hosiery looms.

Woven materials are constructed of longitudinal or warp threads (kette, zettel, aufzug, schweif; organsin; chaine), and transverse or weft threads (schuss, einschuss, eintrag, einschlag; trame), in such a manner that the latter pass regularly at right angles backwards and forwards through the warp, and therefore must turn back at the edge or selvedge (sahlband, sahlleite; lisiére) of the material — which is frequently made of a different or stronger class of warps in order to improve the appearance and strength of the goods.

Owing to the manifold changes of relative position that can be given to the warp and weft threads, a great variety of fabrics can be manufactured. These are classified into the following four groups:—

1. Plain or Smooth Weavings.

The weft thread has only two positions, this being the simplest form of combination possible.

(*a*) *Canvas Fabrics, and the like.*—In these the even warp threads lie above during the first passage of the weft, and the odd threads below, the position being reversed in the second passage.

(*b*) *Gauze-like Plain Fabrics.*—These show the one half of the warp threads, *e.g.*, the even numbers, always above, and the odd ones always under the weft.

2. Twilled Fabrics (Köper; Croisé).

The shott (weft) thread always takes up more than two positions; it may pass over two, three and more warp threads, and produces on the surface of the material diagonal lines, either continuous or interrupted.

(*a*) *True Twill* (*Croisé*).—The lowest form is the 3-shaft twill, in which the weft passes under two and over one of the warp threads alternately. In 5-shaft twills four warps lie above, and one warp under, the weft. If the warps above and below the weft are equal in number, the product is known as double or Batavia twill.

(*b*) *Atlas.*—The warp lies very free, whereby the well-known, smooth surface of the atlas twill—which is highly lustrous in silk goods—is produced. The true atlas is 8-shaft, the bastard atlas a 5-shaft fabric; in the former the weft passes under seven and in the latter under two warps before covering another.

3. Figured Fabrics.

Figured goods are those in which a design or pattern is reproduced by peculiar manipulations of warp and weft, the latter having always more than two (and generally a large number) of different positions. The portion of the material surrounding the design is known as the "ground," the pattern itself being called the "figure". In order to throw up the design the ground may be either plain, twilled, or variously coloured (taffeta ground, gauze ground, etc.).

4. Velvety Fabrics.

On an ordinary plain ground a hairy cover or pile of upstanding or recumbent fibres may be formed. In the case of imitation (Manchester) velvets the pile is formed from the weft, but in true velvet (silk velvet), from an extra warp thread, the so-called pile warp.

CLASSIFICATION OF WOVEN GOODS.

The largest group is formed by cotton goods, to which various arbitrary designations are given. Among the plain goods are: Cottons (Nos. 20 to 30 yarns), calicoes (Nos. 36 to 42 yarns), chiffon, percale (Nos. 50 to 60 yarns), Scotch batiste, Scotch or Viennese canvas, muslins (Nos. 60 to 80 yarns), Mollinos (strong coarse fabrics from Nos. 14 to 26 yarns), tulle and gauze. Among the twills are: Twill, dimity, English leather, moleskin, molton, twilled nankeen. Atlas fabrics: Satin. The figured cottons are very numerous: Repps, piqués, cotton damask, double cords, etc. Cotton velvet is designated "Manchester".

The linen fabrics are divided into: (1) Plain goods, such as linens, creas, stiffened linens, linings, clothings, coatings, trouserings, linen batiste and linen tulle (loosely woven); (2) Twilled and figured goods: Tickings, drills, bed drills, trouserings, table and towel drills, linen atlas and linen damask (Jacquard linen).

Plain hemp fabrics occur as sailcloth, canvas, hempen cloth, hemp linen, sacking and packing cloths.

Jute goods are brought on the market, in ordinary qualities, as packing cloths, sackings and sailcloth, and in finer grades, as carpets, table and bed covers, and curtains.

Woollens are subdivided according to the nature of the raw material, into carded woollens and combed woollens (worsteds). To the first named belong the unfulled or only slightly fulled fabrics, such as flannels, buckskins, ratiné, floconné, half-wool llama or union (warp cotton, weft wool), and also the fulled fabrics, the chief representative of which is known as "cloth," by which term is understood a plain weaving with its surface converted into a smooth felt-like covering by the processes of fulling, roughing, shearing and pressing. This felting is facilitated by using warp threads made with a right-turn, along with left-turn weft yarns.

The combed woollens are also a numerous class. To the plain goods belong: Orleans, wool muslin, crape, mohair, repps. To the twills: Merino, Thibet, cashmere, zanella, serges and lastings. The figured goods include: wool- or upholsterer's-damask, vestings, trouserings, shawls, facings and carpets. Velvety fabrics are wool velvet, wool plush, upholstery plush, Brussels carpeting, astrachan and crimeas.

Equally varied are the descriptions of silks, due to rapid changes of fashion. The most important plain silk fabric is taffeta, which is used for dress goods or for linings. A closely woven taffeta of very strong threads covered with a kind of regular grain, or ribbed, is known by the names of Gros de Naples, Gros de Tours and Moiré silk. Foulard is used for handkerchiefs, but also for dress stuffs as well. Gauze, crape and barège are also included in this class, the latter being a very light translucent dress stuff. Among the twills are chiefly the following: Croisé, serge and atlas, or satin. Figured silks comprise silk damask, broché, gold and silver brocade (interwoven with gold or silver threads). Among the velvety fabrics should be mentioned the true silk velvet which is met with in commerce, both with cut and uncut pile, and is used as expensive dress material and as costly upholstery and curtain goods, etc.

In unions or mixed goods the warp and weft threads are of different materials. In half-silk, for example, the warp is of silk (organzine), and the weft of worsted, mohair, alpaca, or even cotton. Thus poplin contains fine combed wool; semi-taffeta, cotton; and atlas twills very low qualities of cottons as wefts. Half-silk velvet is made on a ground of cotton. Barège is a gauzy fabric with a raw silk warp and combed wool weft.

The number of semi-woollen fabrics is very considerable. For the most part they consist of wool and cotton, very rarely of wool and silk, the warp in many cloth-like fabrics being of cotton and the weft of carded wool, *e.g.*, flannels (plain or twilled, slightly roughed or cropped and lightly fulled); cassinets (twilled or atlas twill, only hot pressed); unions (half-wool llama), half-wool moleskins and half-wool twills.

The half-wool fabrics from combed wool are numerous, among the plain goods being wool muslin (loose, soft tissue), Orleans (cotton twist warp), repp (mostly cotton or carded wool warp with combed wool weft), mohair. Twilled are: half-wool cashmere (silk warp, merino weft), half-merino (3-shaft twill with cotton warp and combed wool weft, or *vice versâ*), zanella (cotton twist warp, wool weft); and the figured and gauzy fabrics comprise half-wool damask (cotton twist warp), and half-wool barège (cotton warp, combed wool weft).

In the half-linen (union) fabrics are distinguished: union canvas (made partly from hemp alone and partly from hemp warp and flax weft), tow canvas, half-tow linens, tow weft sheetings (flax warp, tow weft), half-cotton linens or unions (cotton warp, flax weft, or *vice versâ*). Among the figured goods are union drills (cotton warp, linen weft) and union damasks.

Materials containing threads of other kinds for the production of special effects are not regarded as mixed goods.

THE TESTING OF FABRICS

comprises four groups of investigation, the first of which belongs to the domain of mechanical technology and includes :—

1. Identification of the *mode of weaving*; adjustment of warp and weft ; combination ; external appearance of the individual threads ; yarn number ; doubling twist ; colour, etc.

2. Testing the *breaking strain and elasticity* by the dynamometer. Tearing by hand is an inaccurate test.

3. Determining the "*count*" *of warp and weft* (porter and pick or shott).

4. Determining the *shrinkage or contraction*.

The second group deals with the identification of the raw material :—

5. Examination of the *constituents* of the warp and weft threads ; weighing the fabric.

The third group is concerned with the substances and alterations introduced into the goods by the processes of dyeing and finishing :—

6. Determination of the *finish and dressing material*.

7. Determination of the *waterproof capacity of the fabric*.

8. Determination of the *absorption of moisture*; artificial weighting.

9. *Fastness of the dye* under the influence of weather, dirt, dust, washing, perspiration, ironing, etc.

Finally, in the fourth group are determined :—

10. *The length of the piece* of cloth, etc.

11. *The mordants and dyes employed*.

12. *The presence and amount of arsenic* (if any).

I. DETERMINATION OF THE MODE OF WEAVING, DISTINCTION AND COMBINATION OF WARP AND WEFT THREADS.

In many fabrics the differentiation of warp and weft threads is an easy matter, whereas in others certain indications are necessary. When the sample contains a piece of the selvedge this shows which is the warp, because a selvedge can only be in a longitudinal direction in the stuff. In the case of cloth, buckskin, flannel, etc., and with fabrics generally that have been fulled, raised and cropped, the direction of the hairs composing the pile affords a guide, these being always laid parallel to the warp. If a number of doubled threads are found in a fabric, the others being simple, then the former are the warps. When cotton threads are detected as running in one direction and woollen ones in the other, the cotton yarns usually form the warp and the wool the weft. Warp and weft may also be distinguished by the twist of the yarn, the former being the more tightly twisted of the two. Where the members of one set of threads are equidistant and the others at irregular intervals, the former are usually the warps. In stiffened or starched goods, if only the threads running in one direction can be seen, they may be assumed to be the warp, and if one set appears stiffer and straighter, the other being rough, crooked or crumpled, the former may be regarded as composing the warp and the latter the weft. The *material* also affords a clue, since if one set of the threads is of better and longer material and higher yarn number than the other set, the finer constitutes the warp and the commoner, thicker yarn the weft. Finally, the direction of the twist in the threads is conclusive, so that if one set has a strong right-twist and the other a left-twist, the first may be regarded with certainty as the warp.

After the distinction of the warp and weft yarns has been

effected in one way or another, it then becomes a question, especially for manufacturers, of determining the combination of the warps and wefts. To this end the sample is examined by the aid of a hand glass or thread counter and a strong style or needle, with which the threads can be counted and moved aside, and the cloth analysed. The position of the threads is called their "combination," and is classified as taffeta, twill, atlas, serge, Batavia, and so forth. The result of the examination is plotted on cartridge paper, the projecting warp threads being indicated by filling up the corresponding squares and leaving those referring to the prominent weft threads blank. In this way the weaving pattern of the sample is obtained, and serves as a guide to the weaver in making the stuff, as well as for the preparation of the pattern cards for the loom.

Finally, the individual threads are tested for strength, twist and dye, the methods for which operations have already been detailed.

II. TESTING THE STRENGTH AND ELASTICITY OF A FABRIC.

The plan, so often adopted, of testing the strength of a cloth by tearing it by the hands is altogether unreliable, and rather leads to self-deception, because tearing frequently requires only a certain skilled knack whereby the best material can be pulled in two, and though an experienced man may be able to distinguish a good cloth from a bad one in this way, yet it is impossible for him to determine slight differences in quality, since after he has exerted his strength over a few tests the capacity to distinguish the actual force required disappears.

The sole means of determining the strength of a sample without possibility of error is by means of mechanical dynamometers, which moreover possess the advantage of requiring no skilled knowledge in their application, besides expressing the tensile strength of the sample in terms of weight. This latter faculty in particular is valuable to the manufacturer by enabling him to accurately compare his various products with those of his competitors, whereby he receives assistance in improving his fabrics. The tests will facilitate the detection of defects of manufacture which would otherwise be overlooked, and will undoubtedly contribute also to strengthening the market, since it is only by means of such tests that one can determine what goods are to be considered as durable and strong. The value of these tests is sufficiently proved by the fact that all Army Clothing Departments and other official departments already make the reception of their supplies of cloth, drills, canvas, etc., dependent on their passing definite tests for strength.

Breaking tests also afford the most certain proof to bleachers of cotton and linen goods whether the bleaching process adopted is a rational one, *i.e.*, whether the goods have been weakened or not; and by the same tests an indication can be quickly obtained as to whether body and household linen has been—as is so frequently asserted—improperly treated in the laundry.

In turning to account the results obtained with the testers about to be described, the length of the sample examined, its weight and the degree of force (weight) employed, give the length of the stuff which if suspended would break under its own weight—the width being of course considered as equal in both cases. So, for example, if 15 *cm.* (6 in.) of stuff weigh 1 kilo. (2·2 lbs.), and break under a strain of 50 kilos. (110 lbs.), then the tearing length of the stuff will be

$$R = \frac{0.15 \times 50}{0.001} = 7500 \text{ metres (8202·225 yds.)}.$$

If the strip of cloth is found to have stretched, say 15

TESTING THE STRENGTH AND ELASTICITY OF A FABRIC.

mm. (0·6 in.) at breaking point, then the stuff is said to have a "tenacity" of (in this instance) 10 per cent. of the original measurement.

In addition to the two definitions, expressed by the terms "specific strength" and "breaking stretch," is associated a third, "specific breaking tension," denoting the product of the tension exerted (weight) and the distance traversed (length) referred to 1 gram (0·035 oz.) of the fabric tested.

The subjoined table, compiled by Hoyer,[1] gives, for guidance, the results of several experiments on the count and strength of cloth :—

Fabric.	Weight in Grams per square metre.[2]	Number of Threads in 25 *mm.* (1 in.).		Breaking Strain of a Sample 10 *cm.* (4 in.) wide in kilos. (2·2 lbs.).	
		Warp.	Weft.	Warp.	Weft.
Cotton, unbleached, for shirtings	129-135	56	53	90	74
Cotton, blue, for linings	110-170	60	51	58	50
Linen { Shirting (bleached)	225-235	32	28	73	67
Linen { Lining (unbleached)	205-215	26	22	105	100
Linen { Summer trousering (bleached)	240-260	33	30	230	144
Linen twill for trousering (unbleached)	320-340	30	29	230	200
Orleans, black (cotton warp, wool weft)	63-69	72	62	36	30
Military cloth	80-85	37	37	57	52
Coating cloth (shoddy)	—	30	25	35	34
Sheep's wool cloth { Blouses	60-64	46	43	67	53
Sheep's wool cloth { Lining	45-49	42	37	54	40
Worsted (black)	50	60	55	100	92
Silk (ordinary plain)	20	—	—	120	110

[1] Dammer's *Lexikon der Verfälschungen*, p. 333.

[2] The weight per sq. yd. in oz. is obtained by dividing the weight per sq. metre in grams by $1·893 \times 31 = 33·88$, *e.g.*, $130 \text{ grams} \div 33·88 = 3·8$ oz. per sq. yd.

BREAKING STRAIN TESTERS (DYNAMOMETERS).

Of the various testing machines manufactured, those described below are given as having proved themselves useful and made their way.

(a) *Rehse's Tester* is a pocket instrument which can be recommended when rapid and, at the same time, accurate testing is in question. The stuff is stretched, by means of a pressure screw, between a concave disc and a tube, in such a manner that the entire surface of the tube is covered by the cloth. The sample is pressed upon by a die, driven by a spiral spring, until the stuff, which is forced into the hollow of the concave disc, is torn. The force exerted by the spring is produced by moving a sliding case or drum, the progress of which indicates on a scale the breaking tension and elasticity of the sample under examination.

(b) *Leuner's (Dresden) Tester.*—This apparatus registers the elasticity and strain automatically on a strip of paper. The sample is held by a pair of clamps, which are then attached on the one hand to a fixed hook E (Fig. 63), and on the other to the balance, by suitable studs. It is necessary that the initial distance between the two clamps should always be constant, and that—whether in consequence of improper adjustment or defective connection with the rest of the apparatus—they should not exert a distorting force on the sample. To this end an even tension longitudinal to the threads is essential, and this is ensured by the arrangement of the apparatus. The indicator recording the tension and elasticity consists of a pencil C and a cylinder B, the latter being mounted loose on the draught rod connecting the spring with the clamp, and receiving a rotary motion by means of cone wheel gearing, which motion corresponds to the movement of the clamp at A, *i.e.*, to the total stretch of the cloth. For indicating the amount of extension due to

TESTING THE STRENGTH AND ELASTICITY OF A FABRIC. 129

the action of the draught screw at the further end of the spring, in contradistinction to that attributable to the elasticity of the cloth, the pencil C moves vertically along the cylinder for that distance, the compounding of these two movements resulting in a curve, the ordinates of which denote the tension and the abscisse the elasticity. By placing the spiral screw in its normal position with regard to the cylinder, the former are described by the pencil, and the latter result from the rotary movement of the cylinder.

For convenience in reading the diagram a glass measuring plate engraved on the under side is employed. The

FIG. 63. Leuner's cloth tester.

divisions marked on the plate show the elasticity in percentages direct and the tension in units of weight (grams), a method which dispenses with the necessity for drawing the ordinates. The perpendicular, graduated in weight units, is applied to the outermost point of the curve, and, the line corresponding to the abscisse being applied to the starting point, the result is read off.

When the breaking point is reached the sudden recoil of the spring is prevented by means of a couple of detents. If the tension of the spring is not great it may be released, after disengaging the detents, by controlling the reverse movement of the cylinder B by the hand; otherwise the detents are left in position, and the spring is restored to its normal tension by means of the hand wheel and screw.

(c) *Breaking Strain and Elasticity Tester.*—The sample is evenly fastened between the cheeks C^1 and C^2 (Fig. 64), the exact distance between them—which should be, say, for example, 400 *mm.* (16 in.)—being ascertained at the outset

Fig. 64. Breaking strain and elasticity tester.

by reference to the graduations on the guide rod E. The cloth is then strained by turning the handle F until it begins to tear, the tension being registered on the dial plate in units of weight (*e.g.*, kilos.). The detents G prevent the pointer from receding, and the elasticity is ascertained by the difference between the initial and final length of the sample as measured on the guide rod E, e.g., initial length 400, final 430 $mm.$ = elasticity 30 $mm.$

FIG. 65. Tarnagrocki's cloth tester.

To re-adjust the spring balance to zero, the wheel F is reversed until the cheeks C^1 C^2 can be connected by the piece

H, the spring being then drawn on slightly to loosen the detent G, and afterwards allowed to recoil by continuing to reverse the wheel F.

(*d*) *Breaking Strain Tester* (*by Tarnagrocki of Essen*).—This apparatus, which has lately been adopted by the German Official Clothing Department, is made entirely of metal, and consists of an instrument for measuring the tensile strength of cloth, an arrangement for measuring the elasticity, and an arrangement for absorbing the vibration of the spring. The tension applied at any moment is automatically indicated by a pointer, and can be read off at any time. The vertical form of the apparatus enables the accuracy of the scale to be checked by direct loading with weights. When the sample is fixed in position the wedges must be pressed slightly in order to keep the cloth secure and safe. The tension is produced by turning a hand wheel, the motion of which is communicated by screw gearing to the spindle contained in the pillar S, which spindle depresses the draught rod T connected with the lower clamp. By disconnecting the screw gearing and connecting the cone wheels K and K', as is shown in Fig. 65, the lower clamp can be quickly re-adjusted to its original, or any other desired, position, when the sample has been broken. For convenience of reading, the dial plate is placed on a level with the eye, and the maximum movement of the pointer is recorded by a supplementary hand. In setting up the machine care must be taken to have it vertical and fixed on a firm foundation.

The method of applying the test is as follows : When the sample strip of cloth has been inserted sideways between the wedges of the stretcher m a slight pull on them is sufficient to fix it securely in place. Tension is then applied by turning the hand wheel, and the pendulum P leaves its vertical situation to take up an inclined position until breaking is effected, its fall at that instant being prevented by

TESTING THE STRENGTH AND ELASTICITY OF A FABRIC. 133

detents dropping into a toothed rod in the lower part of the apparatus. The detents are released on lifting the pendulum by the handle *H* and giving a slight pull on the cord, and it can then be lowered to the perpendicular.

In order to ascertain the elasticity of the sample, a meter

FIG. 66. Tarnagrocki's cloth tester.

for this purpose is affixed to the upper stretcher head *m* and rests on a coiled spring, its fine needle points *i i* being pressed into the sample, which should be 30 or 36 *cm.* (12 or 14·4 in.) long and 5 *cm.* (2 in.) wide.

The lighter machine of the same maker (Fig. 66) varies slightly from the foregoing, as will be apparent from the illustration. The tension, produced by turning the hand wheel r, is transmitted by the gearing k to the vertical spindle, and thence to the test strip, which latter may in this machine be reduced to 18 cm. ($7\frac{1}{5}$ in.) in length by 2·5 cm. (1 in.) in width.

In an older model the test sample is fixed horizontally between two clamps. The tension is produced by the spoke wheel h, whereby the counterpoise y is correspondingly moved, as are also the pointer m and the friction pointer n

FIG. 67. Horizontal cloth tester.

on the scale. Directly the cloth breaks, the counterpoise is restrained so that the breaking strain can be read off on the scale. The counterpoise can then be lowered by releasing the detents by a pull on the cord f, and by the winding arrangement c.

The foregoing horizontal machine serves for testing all kinds of driving belts, girthings, strong fabrics, thin ropes, etc. The test sample requires to be from 25 to 50 cm. (10 to 20 in.) in length and 5 cm. (2 in.) in width.

In the official instructions issued to the German Army Clothing Department[1] the testing machine first mentioned is described in detail as follows :—

1. The Dynamometer.

The visible parts of the dynamometer are :—

(a) A case, of horse-shoe form, constituting the front of the instrument, and covered with an engraved dial plate of brass;

(b) A cast-iron frame;

(c) Two rods, each fitted with a clamp and hook—inside the frame—the first being attached to the measuring spring, referred to below, and movable only as far as this spring permits, whilst the other rod can be moved up and down freely;

(d) A large screw fitting on the under side of the frame and worked from the front of the instrument by a hand crank, the female screw being contained in the lower part of the second rod.

The most important part of the apparatus, however, is :—

(e) An elliptical spring, consisting of two blades, and connected to a pointer.

The function of the horse-shoe case is mainly to hold fast one of the points of the spring; that of the frame to keep the clamp rods in position; that of the clamps to hold the sample of cloth to be tested; that of the large screw to apply tension to the object of the test until breakage is produced; and that of the spring to measure such tension.

The amount of force exerted is indicated by a pointer mounted on the axis of a cog wheel below the dial plate and moved forward from zero by the draught exerted by a hook fixed on the clamp rod, and actuating a rack which engages in the small cog wheel.

[1] *Dienstanweisung für die Corps-Bekleidungsämter*, p. 105.

The effect of the motion imparted by the larger screw is to draw the second clamp away from the first, thereby stretching the test sample and drawing on the upper clamp, tightening the spring and moving the pointer.

When the cloth commences to tear, the test is finished, and the pointer retains the position it held at the moment of breaking, its steadiness being explained by the fact that as the spring recoils the clamp hook is released from the rack.

The maximum position attained by the pointer is— expressed in units of weight (kilos.)—an indication of the force necessary to overcome the tension of the sample, *i.e.*, of the breaking strain of the material tested.

2. APPARATUS FOR MEASURING ELASTICITY.

As the stretching of the material is a necessary accompaniment of the breaking strain test, all that is necessary for the determination of the elasticity is an arrangement for recording it.

This consists of a scale, graduated in millimetres, affixed to the cast-iron frame, to which it is screwed loosely, and, being notched, is pushed forward by a stud projecting from the first clamp, with which it is always on a level, the pointer being formed by a small bent rod on the second clamp.

3. APPLIANCE FOR ABSORBING OSCILLATIONS OF THE SPRING.

The following parts belong to the instrument, but are not necessary to the performance of the test:—

(*a*) A quadrangle brass frame about the centre of the springs;

(*b*) A partly toothed brass rod under the horse-shoe case; with

(*c*) A projecting arm or angle piece at the side of the case;

TESTING THE STRENGTH AND ELASTICITY OF A FABRIC.

(*d*) A loose rod which, passing through the case and the brass frame, is pressed by the aforesaid arm against the spring;

(*e*) A spiral spring along the toothed rod to which it is attached by one end, the other being held by a stay thereon; and, lastly,

(*f*) A set of toothed gearing, engaging in the rack work on the brass rod, and consisting of a segmental rack with toothed driving gear (axis in the under frame), brake wheel with brake cone, and fly wheel.

These parts are only destined to take up and absorb the oscillations of the spring at the moment the test is finished, so as to protect the instrument from injurious jarring. This task is performed in the following manner:—

During its tension the spring assumes an elongated form whereby the rod accompanying the spring and pressed by the arm acquires a certain play, and as soon as this occurs the extended spiral spring follows its natural tendency to contract, and so puts in motion the wheel gearing by drawing the rack engaging with the said gear.

After the test object is broken the main spring rises rapidly into its original position, drawing the rod with it, the result being that the arm falls back, the spiral spring is again extended, and the wheel gear reversed. In this retreat the oscillations of the spring are taken up from the rod and carried forward through all the parts to find an outlet in the rapidly revolving fly wheel, and so are rendered harmless towards the apparatus.

The following instructions concerning the handling and use of the dynamometer are also given:—

Handling.—The dynamometer should be handled with care, and must be kept in a dry place, besides being cleaned and oiled sufficiently to ensure its preservation and working easily when put into use.

The best lubricant for the large screw, as well as the pointer and wheel gear, is neat's foot oil, whilst rape oil is better for preserving the spring. A small hole for oiling the screw is bored through the front of the cast-iron frame.

Other precautions necessary for the preservation of the machine will be apparent from the subjoined observations:—

If the dynamometer is subjected to a heavy jar from a blow or shock, then the spring, dial plate, or pointer may become loosened or shifted—even though such be not outwardly apparent—and therefore inaccurate.

Although the spring, which is constructed to bear a load of 500 kilos. ($\frac{1}{2}$ ton), is unaffected by high tension in ordinary measurements, it suffers if left for a long time under very high tension. Moreover, it will seldom stand extension at a temperature below freezing-point, and still less a strong tension resulting from the rotation of the fly wheel; in the former case it breaks easily, and in the latter may be forced out of its bearings.

Finally, it is advisable, for the protection of the spring, not to load it above 250 kilos. (or 5 cwt.).

Application.—The sample to be tested should be of a certain length for determining the elasticity and a definite width for ascertaining the breaking strain of the material, which sample may be taken from any convenient part of the cloth, but must be cut either in the direction of the weft or the warp.

One end of the cutting is fixed in the one clamp and the other inserted in the second clamp, the correct distance between the two clamps being adjusted by the hand before the second one is screwed up.

This accomplished, the screw is rotated at suitable speed (about 100 turns a minute) by turning the hand crank until the sample begins to tear and the pointer remains at rest on the dial plate. In the case of many stuffs, the continued

TESTING THE STRENGTH AND ELASTICITY OF A FABRIC. 139

movement of the hand crank, after tearing has begun, results in a further extension of the material before complete separation is effected, which, however, cannot be regarded as increased elasticity.

The manner of reading off the elasticity and breaking strain has already been indicated. Nevertheless, it should be remembered that the degrees of elasticity are not expressed in millimetres as recorded on the scale, but as centimetres, each fraction being reckoned as a whole number, because it may easily happen that in adjusting the sample by the hand the stuff is stretched rather more than is absolutely necessary.

Before proceeding to a new test, the pointer must be readjusted opposite the zero mark on the dial, and the brass scale on the iron frame pushed back until its zero point is exactly on a level with the first clamp.

After the machine has been lying idle for a long time it is advisable, before commencing the tests, to gradually stretch some solid body, such as zinc plate, several times (say, twice under a load of 200 kilos. and thrice under 80 kilos. tension) in the machine, in order to restore the spring to its normal length, since, when at rest, it is inclined to alter (contract). It may be assumed that under a constant temperature a not unimportant alteration of this nature may occur within a few hours, and in a much shorter time under sudden fluctuations in the weather.

When frost prevails the tests should be performed in a heated apartment, as the machine cannot yield accurate results when under the influence of cold. As a rule, the room temperature is regulated to about 15° R. (68° F.), and the testing commenced only after the machine has been exposed to that temperature for one or two hours.

From the results obtained with the test sample the breaking strain and elasticity of the material are calculated to the total length (warp) and width (weft) of the piece of cloth in question.

In accordance with the conditions of the specification, a piece of the cloth 9 *cm.* (3·6 in.) wide and at least 38 *cm.* (15 in.) long is measured off on the piece by the aid of a superimposed rule, and, after making an incision at both sides, torn off by hand, then doubled over in the width, and fastened in the machine, the distance between the clamps being adjusted to 30 *cm.* (12 in.).

Accurate measurement of the width is a prime essential condition, since the breaking strain increases with the size of the section, and hence every thread has a definite value. It should, moreover, be recalled that irregularities in adjusting the material in the machine or in turning the crank may cause inaccuracy in the results of the test.

The officials are also instructed to check the breaking strain results, given by the dynamometer, by the aid of a spring balance, the regulations prescribing :—

1. The spring balance is attached in the same manner as a sample of cloth, so that the zero point on its dial is next the lower clamp of the dynamometer, in order to facilitate reading off the power employed. It is also necessary that the balance should be carefully adjusted and screwed up.

Up to a tension of 40 kilos. (88 lbs.) the position of the balance can be disregarded, but from this point onwards it must be perfectly level, and the box should not touch either the iron frame or the large screw; this is most conveniently controlled by drawing a sheet of paper lengthwise between the box and the screw. If there is contact at any point the evil may be easily remedied by adjusting the spring balance, or by screwing it up again, or by suitably bending the brasses until the balance assumes its proper position.

2. Each testing performed at medium temperature (68° F.) must be preceded by a few determinations made with

strips of cloth in the spring balance itself, but in order to avoid straining the spring of the latter it should not be exposed to a higher tension than 100 kilos. (220 lbs.).

3. Only the results of the dynamometer between 40 and 70 kilos. (88 and 154 lbs.) need be tested, differences manifested in more extended tests being considered as unimportant.

4. If the results of three to five consecutive tests with the dynamometer agree in each case with those of the spring balance, or, on the other hand, vary among themselves, or are concordant but disagree with those of the balance, then the dynamometer must in the first instance be looked upon as correct, in the second as defective and unreliable, and in the third as inexact but still usable. In the latter event, if the difference cannot be adjusted it must be determined exactly and taken into account in judging the results of the tests.

5. The spring balance itself must be subjected to a preliminary examination at the prescribed temperature by loading it several times with weights of 40, 50, 60 and 70 kilos. (88, 110, 132 and 154 lbs.). If the balance does not weigh true it should be sent to a skilled mechanic for examination.

III. ASCERTAINING THE COUNT OF WARP AND WEFT THREADS IN A FABRIC.

Every fabric must contain a certain count of warp and weft threads, *i.e.*, a definite number within a certain space for each strength of yarn (yarn number) employed. The closeness or density of the texture of the material is considered as falsified when less than the requisite number of warps is present in a given width, or when the number of wefts (picks or shotts) per yard or metre is changed, or again when the prescribed yarn number is replaced by a lower one.

In order to ensure the even distribution of the warps the weaver employs a "reed," which also serves to beat the wefts, the distribution of the weft threads being controlled by the regulator affixed to the loom, so that by means of these two appliances any determined closeness of texture or density can be produced in the fabric.

(a) *Determining the Weft Count.*—This is effected by counting the weft threads by the aid of a lens, the base plate of which is pierced by an opening of definite size. The English standard (shott) is 1 in., and the metric standard 1 cm. (0·3937 in.).[1]

(b) *Determining the Warp Count.*—The same procedure is adopted, but the standard is different, the English "porter" measuring $\frac{15}{16}$ths of an inch. On the Continent, although attempts have been made in the technical schools to introduce the metric standard of 1 centimetre or 1· decimetre (3·937 in.), the count is in many factories still reckoned on old-established standards.

So, for example, in Saxony the count of clothings, upholstery goods, worsteds, etc., is based on the "gängen" in a space of 6 Leipzig zoll ($\frac{1}{4}$ ell, or 5·65 in.), 40 warp threads constituting a "gang"; thus, "5 gang" cloth means that 5 × 40 warps are present per 6 Leipzig zoll (14·12 cm. or 5·65 in.).

Silks are counted in Crefeld and Elberfeld by "fein," which express the number of times 100 reed splits contained in 40 French "pouces" (or 108·4 cm. = 42½ in.).

In France and Switzerland the count is based on the number of threads per French "pouce" (= 2·71 cm. or 1·062 in.) or per 1 cm. (0·3937 in.).

The "buckskin" makers give the total number of threads

---

[1] *Translator's Note.*—To determine the count of cloth for customs duty in France, where the tariff is based on the number of threads in a space of 5 sq. cm., it is advisable to employ a stencil cut to a square of 5 cm. side.

present in the full width (about 55 in.) of the goods. The cloth has therefore to be woven somewhat wider than this in order to allow for the shrinkage in fulling, closely woven and thicker stuffs having to be wider in the unfinished state than more open qualities and light summer goods. Winter goods may be 1·80 to 2·15 $m.$ (70¾ to 84¾ in.), and summer goods 1·52 to 1·80 $m.$ (58 to 70¾ in.) wide.

IV. DETERMINATION OF SHRINKAGE.

Shrinkage exerts an important influence on the value of a fabric. Its extent is determined by pouring hot water over a piece of cloth 50 $cm.$ (19·66 in.) by about 30 $cm.$ (12 in.) wide, and leaving the latter immersed overnight, to be afterwards dried at a moderate temperature without stretching, the ensuing decrease in length giving the shrinkage, which is usually expressed in percentages.

Military cloths should not shrink more than 5½ per cent. in the length (5·5 $cm.$ per 1 $m.$) and 4·2 $cm.$ (1⅝ in.) in the total width of the piece.

Other goods, which are also manufactured for army purposes, generally shrink as follows:—

Drills	- 4 to 5 $cm.$ (p.c.) in the length	and	1 to 2 $cm.$ (0·4 to 0·8 in.) in the width.
Linings	- 8 to 10 ,,	,,	3 to 4 ,, (1·2 to 1·6 in.) ,,
Fustian	- 2 to 3 ,,	,,	1 to 2 ,, (0·4 to 0·8 in.) ,,
Flannel	- 8 to 9 ,,	,,	4 to 5 ,, (1·6 to 2 in.) ,,
Calico for Invalids' Shirts	- 3 to 4 ,,	,,	1 to 2 ,, (0·4 to 0·8 in.) ,,
Calico for Drawers	- 1 to 2 ,,	,,	[$nil.$]

DETERMINING THE THICKNESS OF CLOTH.

In order to be independent of the (unreliable) estimation of thickness of a cloth by the "feel," an apparatus consisting of a pair of discs connected by a micrometer screw graduated in millimetres (0·03937 in.) is employed.

V. EXAMINING THE CONSTITUENTS OF THE WARP AND WEFT. WEIGHING.

A piece of the cloth about 1 in. square is cut out and dissected into warps and wefts. The piece must be large enough to contain specimens of all the different kinds of yarn present in the material, and all doubled threads must be untwisted and the individual threads examined.

The further examination is conducted in the manner prescribed in a previous section, the dye or dressing being removed by boiling in water, dilute acetic or hydrochloric acid, or dilute caustic alkali, etc., and the preparation afterwards made submitted to microscopic examination, with or without the assistance of various solutions employed for staining or moistening the fibre.

WEIGHING ON THE BALANCE.

For buyers and sellers of goods, who—when, for example, travelling—require to ascertain the weight of an article quickly, a pocket balance has been constructed by Zetsche of Gera, which will indicate the weight per square yard or square metre of a small sample cut by the aid of the stencil appertaining to the instrument. A second stencil enables the weight of extra heavy cloth to be determined. This balance saves the tedious operation of calculating the weight from the count of warp and weft and the yarn numbers. It is also constructed for use as a stand balance with two stencils and separate scales to show the weight in grams per square yard or square metre.

DETERMINING THE WEIGHT OF CLOTH FROM THE COUNT AND THE YARN NUMBER.

The weight of a square metre of cotton cloth is found from the number of warp and weft threads and the (metric) yarn number by the following formula:—

$$G = 60 \frac{K + S}{N}$$

THE CONSTITUENTS OF THE WARP AND WEFT. WEIGHING. 145

G being the weight of 1 square metre of cloth, K the number of warp threads per 1 $cm.$, S that of the wefts per 1 $cm.$, and N the yarn number. If these numbers are different in warp and weft, the arithmetical mean of the two is taken. Thus, for example, the weight of a piece of cotton cloth 36 metres in length and 85 $cm.$ ($33\frac{1}{4}$ in.) wide = 30·6 sq. $m.$, made of No. 20 warp and No. 24 weft, with 22 warp threads and 24 wefts per 1 sq. $cm.$, is :—

$$1 \text{ sq. } m. \text{ weighs } 60 \times \frac{22 \times 24}{22} = 125·4 \text{ grams};$$

therefore, 30·6 metres weigh 3·8 kilos.

The weight will be increased somewhat in the rough cloth by the sizing and in finished goods by the dressing.

The weight of the goods and the count of yarn threads being known, the yarn number is deduced by the formula :—

$$N = \frac{60 \ (K + S)}{G}$$

The weight of linen piece goods is reckoned by the formula :—

$$G = 168 \ \frac{K + S}{N}$$

and, conversely, the yarn number is found from :—

$$N = 168 \ \frac{K + S}{G}$$

DETERMINING THE WEIGHT OF THE INDIVIDUAL CONSTITUENTS OF THE CLOTH. QUANTITATIVE CHEMICAL ANALYSIS OF THE FABRIC.

1. *Mixed Fabrics Containing Wool and Cotton.*

(*a*) *Estimation of Moisture.*—Five grams of the fabric are dried at 100° C. until the weight is constant, the loss indicating the amount of moisture present.

(*b*) *Estimation of Cotton.*—Five grams of the stuff are taken and boiled for ¼ hour with 100 $c.c.$ of 0·1 per cent.

caustic soda solution, then washed with water and treated with lukewarm 10 per cent. potash solution, or caustic potash, until the wool fibres are decomposed; or the liquid is finally raised to boiling heat. Washing with water is next performed and followed by treatment for $\frac{1}{4}$ hour with hydrochloric acid in the water bath; then another washing with water, boiling $\frac{1}{4}$ hour with distilled water, washing with alcohol and ether, and finally drying at 100° C. until the weight is constant. The residue is cotton.

(c) *Estimation of Wool.*—Five grams of the cloth are boiled with 100 *c.c.* of soda solution for $\frac{1}{4}$ hour and washed with water; then steeped for 2 hours in 58° B. sulphuric acid and washed with water (to prevent the fibre from heating). After boiling for another $\frac{1}{4}$ hour with water, wash successively with water, alcohol and ether, and dry at 100° C. until constant. The weight is that of the wool.

(d) *Dressing and Dye* are found by difference.

2. *Mixed Fabrics of Cotton and Silk.*

(a) *Estimation of Moisture:* as above.

(b) *Estimation of Cotton:* as above.

(c) *Estimation of Silk.*—Five grams of the stuff are boiled for $\frac{1}{4}$ hour in 100 *c.c.* of soda solution, washed with water and treated with hydrochloric acid in the water bath. Wash with water and boil $\frac{1}{4}$ hour with distilled water, then boil with alcohol and ether. Dry at 100° C. and weigh. Next steep in boiling zinc chloride solution for an hour, wash with acidified water, then with water, alcohol, and ether, and dry finally at 100° C. The difference between this weight and the preceding one gives the weight of the silk.

The silk may, however, be estimated much more rapidly (*viz.*, in about ten minutes) by means of Löwe's reagent, this liquid dissolving silk very quickly, but having very little action on wool and cotton.

(d) *Dressing and Dye*: as above.

3. *Mixed Fabrics of Wool and Silk.*

(a) Moisture, (b) silk, (c) wool, and (d) dye and dressing, are all determined as above.

4. *Mixed Fabrics of Cotton, Wool and Silk.*

These estimations are all effected as already described.

ANALYSIS OF VARIOUS QUALITIES OF UNBLEACHED COTTON GOODS.[1]

	I.	II.	III.	IV.	V.	VI.
Material:—						
Fibre - - - per cent.	47·29	53·02	60·75	70·84	80·51	81·78
Normal moisture - ,,	4·11	4·61	5·28	6·16	7·02	7·11
Weight of cloth - ,,	51·40	57·63	66·03	77·00	87·53	88·89
Dressing:						
Water - - - ,,	6·01	5·02	4·65	3·07	2·01	2·89
Dressing and fat - ,,	12·77	13·36	13·33	12·43	8·30	3·33
Mineral matter - ,,	29·82	23·99	15·99	7·50	2·16	4·89
Weight of dressing - ,,	48·60	42·37	33·97	23·00	12·47	11·11
The normal moisture being 8 per cent., the excess of moisture is therefore - - - per cent.	2·12	1·63	1·93	1·23	1·03	2·00
Mineral substances:—						
China clay - - - -	much	1·08	11·45		mainly	
Manganese chloride - -	little	2·42	traces	—	—	—
Calcium chloride - - -	—	0·43	2·50	—	—	—
Zinc chloride - - - -	little	6·06	2·04	traces	—	—
Sodium chloride - - -	traces					

[1] Thomson, *Sizing of Cotton Goods* (1877), p. 150.

VI. DETERMINATION OF THE DRESSING.

During the various operations of washing, bleaching, etc., the goods lose in weight, so that the manufacturer is constrained, in order to produce a marketable article, and at the same time indemnify himself for the deficit in weight, to employ a moderate amount of loading. Dressing is therefore not always to be regarded as an adulteration, but is, for the most part, solely an embellishment.

According to Hoyer, cotton goods in the rough should nsist of 83 per cent. fibre, 7 per cent. natural moisture, 8·5 per cent. starch, fat (used in softening the yarn), and about 1·5 per cent. ash. After washing and bleaching, however, only 78 per cent. of fibre is left, and therefore, by the addition of dressing, the finished cloth consists of 78 per cent. fibre, 7 per cent. natural moisture, 7 per cent. starch and 7·5 per cent. mineral matter. The reduction of the percentage of fibre below 78 per cent. in the ordinary condition, or 85 per cent. in the dry state, should therefore be inadmissible.

Linen should be finished without any dressing, or at most with starch (no mineral matter), so that it loses not more than 2 to 5 per cent. when boiled.

Carded yarn goods are not dressed, but contain water by reason of their hygroscopic properties. Sometimes this faculty is increased by the addition of hygroscopic materials. Also in thick stuffs the fulling from shearings is found.

Worsteds are also frequently impregnated with hygroscopic substances.

In the process of silk manufacture the admixture of extraneous substances is practised to a considerable extent, the threads being sometimes loaded to as much as 400 per cent. before weaving.

Accurate classification of the dressing by the appearance

and feel of the goods is scarcely possible, since with the changes of fashion new stuffs appear with a finish for which a special designation must be found in each case. Well-known names for finish are: dull finish, gloss and high lustre finish, filling and covering finish, moiré finish, hard and soft waterproof finish, light, loose and thin finish, in contradistinction to heavy and full finish, moist and linen finish, elastic and stiff finish, and so on.

Various dressing materials are used to produce any determined finish, starch, flour and mineral matters being most frequently employed, especially, on the one hand, to give the goods stiffness and "feel," and, on the other hand, to conceal defects in the cloth, give a solid appearance to goods of open texture, and at the same time improve their weight, the latter practice being often pushed to extremes, for example, in the cotton industry.

When a fabric filled in this manner is placed in water and rubbed between the hands, the dressing is removed, and the quantity employed can be easily determined in this way. The mineral substances used serve chiefly for filling and weighting, and necessitate the simultaneous employment of a certain quantity of starch, flour, glue, or other agglutinant matter. In order that the latter may not render the cloth too stiff and hard, further additions of some emollient such as glycerine, fats, oils, soaps, etc., are necessary. The dressing is coloured blue by means of ultramarine, indigo carmine, etc., and it is also important to add some antiputrefactive or antiseptic substance to prevent the development of fungi and moulds. Wax stearin and paraffin are used as being the only substances capable of developing a high lustre in calendering, pressing, etc.

The largest number and greatest variety of dressings are used for cotton and mixed cotton goods in greater or smaller proportions. For the most part flour, starch and some weighting material predominate, the additions of glue, gelatin, tallow, wax, soap, paraffin, antiseptics, colouring matters, etc., being but small. Acetate and sulphate of lead, alum, china clay, glucose, etc., are used for weighting.

For dressing woollen goods preference is given to Carraghen moss; also glue, gelatin, dextrin, starch, albumin, waterglass (alkali silicate), etc.

Velvet, when dressed at all, is treated with gelatin, gum arabic, tragacanth, etc., on the under side of the cloth.

For silk finishing, flea-wort, tragacanth, gum arabic, shellac, gelatin, etc., are employed.

From these general observations we will proceed to describe the performance of the tests for identifying the dressing materials employed.[1]

1. Physical Examination.

Whether a fabric is finished on one side or impregnated with dressing will be detectable at once, as well as whether it is glazed or merely calendered. By holding the goods against the light starch dressing will be recognised, and such goods if rubbed between the fingers will lose their stiffness. Loading is revealed by the evolution of dust on rubbing; and by the aid of a magnifying glass it can be easily ascertained whether the covering of dressing is merely superficial, or penetrates into the substance of the fabric, and also whether it contains mineral substances or not.

2. Chemical Examination.

(*a*) *Determination of Water.*—A weighed quantity of the fabric is dried at 100° C. until the weight ceases to sensibly decrease. The weighings must be performed in tightly

---

[1] Depierre, *Appretur der Baumwollegewebe*, p. 439.

closed glass vessels, the dried fibres being very hygroscopic. The difference between the first and final weighings gives the amount of water; when this is comparatively large a high degree of weighting may be suspected, since starch absorbs much more water than the pure cotton fibre.

(b) *Determination of the Extraneous Substances.*—A weighed and thoroughly dried piece of the fabric, about 10 sq. ins. in size, is treated at boiling heat with malt extract, washed, dried and weighed, the loss in weight being that of the extraneous matter. As a few insoluble soaps may be left behind, the sample is boiled for a short time in dilute acid, and reweighed after drying.

After this it is necessary to ascertain the nature and quantity of the various dressing materials, to which end the fabric is boiled for several hours in water, whereby starch, gum, and all thickening materials, together with the soluble salts and the earthy bodies, are loosened from the fibre. After removing the fabric the residual liquid is filtered, the filtrate and residue being then examined in detail.

(a) *Examination of the Filtrate.*—The liquid being concentrated, a few drops are taken and tested with dilute tincture of iodine: a blue to red violet coloration denotes starch.

After further concentration from 2 to 3 vols. of alcohol are added, which will precipitate, in addition to a series of metals, glue, dextrin and gum.

Glue is precipitated from its solution in water by a solution of tannin.

Gum and *dextrin* may be distinguished by their polarimeter reading, the former giving a levo-rotation and the latter a dextro-rotation.

When both are present, the aqueous solution on treating with lead acetate deposits gum in the cold, both being precipitated in the warm. If the liquid contain an organic body giving no precipitate in this test, the presence of Carraghen moss, Hai-tao, is concluded.

One portion of the concentrated liquid is warmed on the water bath with a few *c.c.* of hydrochloric acid (the flask being fitted with an inverted condenser), and is then tested for *sugar* with Fehling solution.

For detecting the soluble salts, such as alum, zinc chloride, etc., that may still be present, the ordinary chemical tests (see below) are applied.

One part of the liquid is now evaporated to dryness and treated with acid potassium sulphate. An odour of acrolein indicates the presence of glycerine.

(β) *Examination of the Residue.*—This contains the earthy weighting and filling materials, such as China clay, gypsum, lime, etc., insoluble in water.

(γ) *Examination for Fat and Resin.*—A sample of the fabric is boiled with soda solution and the filtrate precipitated. When fat is present a layer of fatty acid floats on the surface, whilst in the case of resin a precipitate of sylvic acid is formed.

To determine the fat quantitatively a sample of the cloth is treated with water-free ether in a Soxhlet extractor, the evaporation residue being weighed.

For practical purposes the quantitative analysis of dressing mixtures is generally of little importance, since the finisher, when the approximate quantitative composition is known, can easily determine the quantitative ratio of the constituent parts by a few trials.

The *mineral constituents* of the dressing may also be determined from the ash, as follows: A sample of the fabric to be tested is incinerated and calcined in a porcelain crucible and the ash boiled with nitric acid. If effervescence ensues, the presence of carbonates of the earths is indicated. The filtrate is evaporated to dryness on the water bath, the residue dissolved in dilute nitric acid and filtered from the insoluble portion, the liquid being then treated with sulphuretted hydrogen; a black precipitate indicates *lead*.

Confirmation of the presence of this metal is obtained by dissolving the precipitate in nitric acid and adding potassium bichromate.

After filtering off the precipitate produced by sulphuretted hydrogen and removing the excess of the latter by boiling the filtrate, *iron* is sought for by adding ammonia and ammonium sulphide, and the filtrate from any resulting precipitate is tested for *barium*, *calcium* and *magnesium*, to which end the filtrate is treated with hydrochloric acid to decompose the ammonium sulphide and heated to drive off sulphuretted hydrogen, and is then rendered alkaline by ammonia and treated with ammonium chloride and carbonate. Any precipitate formed is filtered, washed, dissolved in dilute hydrochloric acid, and calcium sulphate solution added to the liquid. An immediate precipitate indicates *barium*, whilst the absence of such points to *calcium*, the presence of which can be confirmed by ammonium oxalate.

The filtrate from the precipitate produced by ammonium carbonate, or, if such precipitate be lacking, the solution containing that reagent, is tested with sodium phosphate for the presence of *magnesium*.

Should there be any insoluble residue from the original treatment of the ash with nitric acid, it may contain silica, barium sulphate, tin oxide, gypsum, clay or iron oxide. It is therefore boiled with soda solution, whereby silica is rendered soluble and gypsum is decomposed, and is then filtered. After washing, the deposit is treated with cold dilute hydrochloric acid and tested for iron and calcium in the manner already described. The filtrate is acidified with hydrochloric acid and evaporated to dryness, the residue being then taken up with water and hydrochloric acid. Any insoluble residue of *silica* is filtered off and the liquid tested for *sulphuric acid* by barium chloride.

The residue undecomposed by soda solution, or soluble therein but insoluble in hydrochloric acid, and which may contain barium sulphate (heavy spar), clay or tin oxide, is dissociated by fusing with a ten-fold amount of dry soda in a porcelain crucible, the mass being then treated with sodium bicarbonate and water, and filtered. The washed precipitate is boiled with strong hydrochloric acid, and sulphuretted hydrogen is added to the liquid. A yellow precipitate indicates *tin*. This is filtered off, and one half the filtrate is tested for *alumina* with ammonia, the remainder for *barium* with sulphuric acid. From the filtrate from the fused mass *silica* (proceeding from the decomposition of magnesium silicate and sodium silicate, waterglass) is precipitated by concentration after the addition of hydrochloric acid, the filtrate being tested for *sulphuric acid* by barium chloride.

VII. ESTIMATION OF THE WATERPROOF PROPERTIES OF CLOTH.

The various fabrics, and particularly cotton, linen and hemp goods, are rendered waterproof by coating or impregnating with various substances. So, for example, use is made of solutions of caoutchouc or guttapercha with addition of pitch, resin, linseed oil, varnish or oil; solutions of fat, paraffin, or tar in benzol or carbon bisulphide, and solutions of ferrous sulphate, cupric sulphate, alum, aluminium acetate (with or without subsequent soaping) being employed for impregnation. A number of formulæ for the manufacture of these waterproofings are in existence, some being public property, others protected by letters patent, and others again carefully kept secret.

ESTIMATION OF THE WATERPROOF PROPERTIES OF CLOTH. 155

All that we have to deal with in this place is the testing of the results attained thereby.

The official regulations of the (German) Clothing Department prescribe for sailcloth the following tests: A piece of the cloth 25 sq. *cm.* (4 sq. in.) in size is folded twice, like a paper filter, and placed in a suitable glass funnel, where it is subjected to the pressure of 300 *c.c.* of water. At the end of 24 hours nothing more than a number of equally-

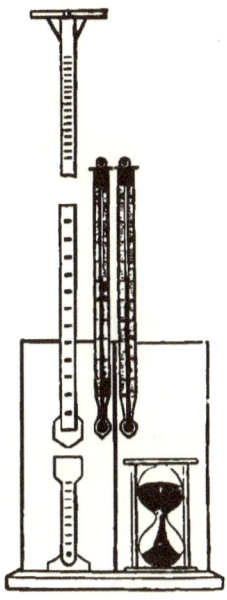

FIG. 68. Gawalowski's waterproof tester.

distributed drops of water should be perceptible on the under surface of the filter, without the stuff being wet through. This test may of course be performed on other classes of goods.

An apparatus for testing the waterproof qualities of cloth has also been constructed, wherein a column of water 12 in. (30·84 *cm.*) high is allowed to act upon the test sample of cloth, the amount of water passing through the sample being collected in a graduated glass placed underneath. The pressure can, of course, be also adjusted in a simple manner to twice the above-named height. This apparatus (Fig. 68), constructed by Gawalowski of Brünn, may be described as follows :—[1]

A quadrilateral glass box with leaded edges is provided with a tight-fitting door in front and a tight-fitting lid pierced by four apertures, and is divided into two parts by a glass partition. An air hole 1 $mm.$ in diameter is bored through the cover on one side, and through another aperture is inserted a burette, a third opening permitting the insertion of a thermometer graduated to fifths or tenths of a degree. The burette, 15 ins. (37 $cm.$) in length, is graduated, and fitted with a cover at the top, the lower extremity being closed by a metal attachment resembling a polarising tube; but instead of a glass disc, as in such tubes, the sample of cloth under examination, cut to the correct size, forms the cover. A slanting outflow is cut through the metal attachment, under which is placed, in the glass case, a small measuring glass, graduated into at least 5 to 7 $cm.$ The funnel placed thereon can be closed by means of a ground glass disc containing a small central air hole.

As is shown in the figure, a second thermometer is passed through the cover, and a twenty-four hour sand glass or other instrument for recording time (fitted, if necessary, with a signalling apparatus) is placed in the glass case.

The apparatus is manipulated in the following manner: A circular disc is cut or punched out of the sample of cloth, and fixed in position in the metal attachment, the passage of water around the edges being prevented by india-rubber rings. The door being then tightly closed, the tube is filled

[1] *Leipziger, Monatschrift,* 1893, p. 221.

with distilled water up to the zero point, and the time required for the transfusion of a certain quantity of water, or the quantity of water passing through within a certain time (twenty-four hours), is noted. If no appreciable quantity of water passes, but a little moisture has collected on the under side of the sample, this will be indicated at the end of twenty-four hours by the differential reading of the two thermometers. The inventor of the apparatus has, in this manner, tested a large number of various samples, and found that many of the articles allowed from 1 to 6 *cm.* of water to pass through after a lapse of five hours, others again in sixteen, eighteen, and twenty-three hours; others again passed so much in five hours that the collecting vessel was filled to overflowing, and finally other samples were so thoroughly waterproof that even after an exposure of some weeks the amount of moisture transfused was so small that nothing beyond slight variations between the readings of the two thermometers was perceptible.

VIII. DETERMINING HYGROSCOPICITY.

Reference has already been made, in treating of the conditioning of yarn, to the importance of estimating moisture. The percentage of moisture can be increased in the case of woollen fabrics to 32 per cent., silks to 27 per cent., cottons to 25 per cent., and linens to 22 per cent., by suitable procedure, such as storing the goods in damp rooms or by using certain hygroscopic dressings, without the goods, however, feeling damp.

To determine such moisture about 5 grams of the fabric are dried in the manner already described—preferably in a weighing bottle—at 100° C. in a drying oven for one to two hours, until the sample ceases to lose weight.

IX. TESTING THE FASTNESS OF THE DYE.

The demands made on the permanence or "fastness" of dyes are manifold. Since, however, absolute permanence is unattainable, the term has to be somewhat limited and qualified with the designation of the influence to be withstood, such as light, air, wear, washing, etc. Dyes for military cloths, which when in wear are exposed during the greater part of each day to the influence of light and air and frequently to rain, must stand different tests as regards fastness, to such as are applied to goods like valuable silks, which are rarely exposed to the sun's rays, are but seldom worn, and then only in an artificial light. In curtains and carpets the capacity to withstand the action of light is the chief essential, whereas in underwear the colours must stand the effect of soap in washing, and in the case of stockings it is necessary that the colours should not come off whilst in wear. No special fastness towards light is demanded of coloured linings, but on the other hand they should not stain in wear, and must be able to resist the action of perspiration, and the same applies to mattress and corset fabrics, etc. One requirement frequently made in respect of dress materials is that the colour shall not fly under the influence of street mud.

In requiring fastness of colour, regard should be had to the material of which the fabric is composed. In the case of shoddy or inferior woollen goods that are only intended to wear for a short time, expensive, permanent colours that would last longer than the cloth itself will not be needed. On the other hand a correspondingly high quality of material and capacity of resistance to light and air are rightly demanded in the case of military cloth, which is exposed to a deal of rough wear.

So far as the dyes themselves are concerned, they can be

determined on the fabric with a greater or smaller degree of facility, the examination necessitating, however, some acquaintance with dye stuffs and methods of dyeing.

Reference is here necessary to a very common error, *viz.*, that the same dye is equally permanent on all fibres. A consideration of the different chemical constitution of the fibres will explain why indigo carmine, for example, is very fast on silk, but not at all so on cotton. Another circumstance of frequent occurrence should also be mentioned, *viz.*, that a fast colou rwhen used in a diluted condition for dyeing light shades is less permanent than when used for dark shades, a good example of which is afforded by the alizarine colours, which are faster on wool than any other dyes, but which are less permanent when used for the production of mode colours than for dyeing darker and richer shades.

The tests for permanence in dyes are applied as follows :—

(a) *Washing Fastness.* — Colours to be proof against washing must be able to stand both the mechanical friction as well as the action of the alkaline liquid and high temperature of the operation. If, under these conditions, the colour remains almost or quite unaltered, and does not stain other coloured or white fabrics washed in contact with it, it is said to be fast under washing.

For the purpose of testing this quality, coloured yarn is plaited with white yarn, or a cutting of the fabric under examination is taken, and immersed in a solution of 5 grams of soap in 1 litre (0·8 oz. per gallon) of distilled water, and pressed therein for two or three minutes at 40° C. (hand temperature), then left for twenty minutes in the solution, rinsed and left for another twenty minutes in the rinsing water, to be finally wrung and dried.

If the colour ought to be particularly fast the soap

solution is heated to 55° C., and the treatment repeated several times over.

This test is applicable to fabrics, whether composed of wool, cotton, or a mixture of both.

(b) *Fastness under Friction.*—Colours on stockings, hosiery yarns, corset stuffs and other fabrics intended to be worn next the skin, must be permanent under friction, and must not rub off, stain or run, *i.e.*, the dyed materials must not give up their colour when worn or in rubbing contact with white or light coloured articles of clothing or the human epidermis.

The test consists in rubbing the material by hand on white—not too smooth—paper, or, better still, on a white, unstarched cotton fabric. In order to obtain reliable, comparable results, the rubbing must be equal in all cases, and friction surfaces of as near as possible the same constitution should be employed.

(c) *Resistance to Perspiration.*—In addition to fastness under friction, power to withstand the action of perspiration is also required, more particularly in stuffs coming in contact with the human skin, and having to absorb the excretions therefrom. This action is intensified by the warmth of the body, by friction, and above all by the fact that the perspiration in the absence of air is obliged to dry with all its constituent matters on the absorbent fibres, and that by the frequent repetition of this process the acids of perspiration (acetic, formic and butyric) become so concentrated that they act destructively on the fibre.

The effect of perspiration on stockings which are repeatedly worn during prolonged journeys on foot, can be estimated. For testing a colour it has been recommended to place a piece of the dyed material on the back of a horse beneath the saddle and examine the effect of a brisk ride, or the test may be performed as follows:—

A bath of dilute acetic acid—containing about 6 *c.c.* of 30 per cent. acetic acid in 1 litre of distilled water—is prepared and warmed to a temperature (37° C.) corresponding to that of the body. In this the sample is dipped and rubbed vigorously with the hand, being then dried, without rinsing, at 20° to 25° C. between parchment paper. This operation is several times repeated, and the more frequently this is done the nearer will the test approximate to actual conditions of wear.

(*d*) *Fastness against Rain.*—This quality is more particularly required in silk materials for umbrella making. The skeins of silk intended for the manufacture of such fabrics are tested by plaiting them with undyed yarns, and left to stand all night in cold, distilled water. The water should not be more than slightly discoloured, whereas the white yarn should not be stained in the least. For woollen yarns this test is occasionally made more stringent; the yarn is plaited with undyed yarn to a queue, and then boiled for ten minutes in water. When wrung and dried the colour should not have deteriorated, nor should the white yarn be stained.

(*e*) *Resistance to Street Mud and Dust.*—This quality is specially exacted for ladies' dress goods, and is tested as follows :—

1. Sprinkling the moistened sample with lime and water, drying and brushing.

2. Sprinkling with a 10 per cent. solution of soda, drying, brushing and noting any change of colour.

3. *Ammonia Test.*—Immersing the fabric in concentrated ammonia for three minutes and observing the colour both in the damp and in the dry state.

4. Ten grams of soda are dissolved in 1 litre of water and mixed with 10 grams of lime—previously slaked and reduced to milk of lime by the addition of water—and 12 *c.c.* of am-

monia. After stirring well up together, the mixture is left to settle, the supernatant liquid poured off, and the residue employed for steeping the sample for five to ten minutes, after which the latter is dried without rinsing, and is finally brushed, any alteration in colour being noted.

(*f*) *Fastness to Weather, Light and Air.*—Every shade of colour succumbs to the influence of the sun, light and air, although in some cases it is only after prolonged exposure that fading becomes noticeable. The degree of permanence can only be determined by exposure to light, to which end one half of the sample is covered with a closely surrounding, but readily movable, paper wrapper, and the whole suspended in the open air in such a position that it is fully exposed to the sun's rays, but sheltered from rain. The object of the paper wrapper is to enable (by removing it at any time) the degree of alteration effected by the exposure to be ascertained. In order to establish a time standard of the fastness to be expected from any dye stuff under these conditions, normal check tests are made with one or two colours of known permanence, *e.g.*, Turkey red or a medium indigo blue on cottons. The samples should be examined daily in order to ascertain the exact time when alteration begins. In the case of Turkey red this will be on the 25th to 30th day, and between the 12th and 15th days for indigo, in summer, or double these periods in winter time. The fastness of other colours can then be estimated in comparison with these.

Attempts have been made to set up standard degrees of fastness, according to which colours that remain without appreciable alteration after an exposure to direct summer sunlight for about a month are classified as "fast," and those undergoing appreciable change under the same conditions as "fairly fast". "Moderately fast" colours are those altering considerably in 14 days; and, finally, those more or less completely faded in this latter term are designated as "fleeting".

A "light-test" apparatus for quick determinations has been devised by Ferd. Victor Kallab of Offenbach. The samples to be tested are suspended vertically in the apparatus and continuously exposed to the sun's rays, the position of the apparatus being changed in conformity with the apparent movement of the solar orb. The action of the rays is strengthened by concentration on a small surface by the aid of a lens 200 $mm.$ (8 in.) in diameter, and with a focal length of 420 $mm.$ ($16\frac{1}{2}$ in.).

Professor von Perger of Vienna proposes a testing apparatus consisting of a plano-convex and a bi-convex lens, the former, with its flat surface turned towards the light, serving to parallelise the rays of an arc lamp, situated at the focal length of the lens, which rays encounter the second lens placed in their path at a suitable distance away. A metal disc placed at a point between the second lens and its focus receives the sample to be tested.

In estimating the capacity of a dye to withstand weather, the country where the material is to be worn must be taken into consideration, since the climate and seasons of various latitudes exert a considerable influence on the rate at which a dye will fade from one and the same material. Thus it is certain that, for example, the colour will be more strongly affected in a given time on the sea coast than in inland districts, and that dark colours are not so durable in southern countries as in northern climes. Permanence is, furthermore, influenced by the material on which the colour is dyed; on poor material, *e.g.*, shoddy, the same degree of fastness cannot, by reason of the price, be expected as in stuff of better quality. Finally, it will be noticed that deep, full colours do not fade so rapidly as light shades.

(*g*) *Resistance to Ironing and Steaming.*—Stuffs, especially for men's wear, which are to come under the hands of the tailor, and corset materials, should not lose their colour when ironed, or, at any rate, the colour should recover its original appearance after a short exposure to the air. This is tested by hot ironing a sample or by drying it on a hot metal plate. In the same manner, capacity to withstand steaming is demanded of many cloths, this latter property being determined by steaming a sample laid between the folds of a larger piece of steamed cloth, during which operation the colour should remain unaltered.

X. MEASURING THE LENGTH OF PIECE GOODS.

For the rapid and accurate measurement of narrow and wide piece goods various machines, some of which fold the

FIG. 69. Cloth measuring machine.

material at the same time, are recommended. The simplest method of measuring is to pass the goods over a roller and read off the length recorded by a counter affixed thereto.

A very useful measuring apparatus is that displayed in Fig. 69. According to the nature of the stuff to be measured, the periphery of each of the measuring wheels is fluted, covered with fish skin, or studded with pins, whereby sufficient adhesion to the material is ensured to keep the latter from slipping by the wheels without moving them. Injurious friction is prevented by counterpoising the main shaft, and the machine is fitted with a universal joint which enables it to be moved in any direction. By an interchange of wheels the machine can be employed for measuring according to any system. In the case of light materials, only one measuring wheel is employed.

XI. DETERMINATION OF MORDANTS AND DYES.

It is frequently desirable to be able to determine the dye stuff by a few tests, the difficulties in the way of such simple determinations being, however, as will readily be understood, considerably increased by the manifold methods employed for fixing the colours on the fibre, and also by the incessantly increasing number of dyes in use. Moreover, most shades are produced by the employment of two or more dye stuffs.

The identification of several dye stuffs present at the same time in any fabric necessitates, on the part of the operator, both a thorough knowledge of the behaviour of dye stuffs and, above all, a considerable acquaintance with the practical operations of dyeing. Even then, the analysis of colours on the fibre is in individual cases almost impossible : *e.g.*, the detection of dyewood colours in presence of certain anthracene dyes, and *vice versâ*.

In many instances the nature of the fibre (whether vegetable or animal) or the nature of the mordant affords a clue to the class of dye employed, and in other cases the *absence* of certain dyes can at least be determined. Thus, for example, the presence of basic coal-tar dyes on cotton and acid coal-tar dyes on wool can always be conjectured. Cotton will frequently be found dyed with benzidine colours, whereas the anthracene colours, with the exception of alizarine, are still seldom used, and need hardly be sought for. In wool the circumstances are diametrically opposite, anthracene colours being more frequently found. When tannin and tartar emetic mordants are detected in half-woollen or half-silk (cotton and silk) goods it is very probable that two dyes of opposite character, one basic and the other acid, have been employed to produce the colour. The application of the fibre, and the quality of the fabric, are points to be considered. Woollen cloths are fulled in the piece, the fibre used for preparing the yarn being dyed loose before spinning, on which account the colours must be very fast in order that they may not bleed or separate from the goods in fulling. As a rule, cheap fabrics are not dyed with exceedingly fast or expensive colours.

In testing, both the natural dye stuffs and the coal-tar dyes must be borne in mind as being almost equally in general use. The natural dye stuffs, such as logwood, fustic, madder, etc., are fixed on animal and vegetable fibres by the aid of mordants, mostly inorganic compounds, such as alumina, tin, copper, iron salts, etc., and most frequently several of these mordants are employed together, because a faster colour is thereby obtainable. On the other hand, the coal-tar dyes, with the exception of the alizarine colours, will fix on animal fibres without, for the most part, any special mordanting being required, the addition of glauber salt, common salt, etc., in wool dyeing, being designed to prevent the too rapid deposition of the colour on the fibre, whereby irregularities of colour are liable to ensue. In the use of alizarine colours a preliminary mordanting is, however, essential. For fixing most coal-tar dyes on cotton, tannin and tartar emetic are used, but in other cases, *e.g.*, the fixing of the (still seldom used) alizarine colours on cotton, other inorganic mordants are employed.

The mineral colours—chrome yellow, orange chrome, iron buff, yellow, manganese bistre and Berlin blue—are, with the exception of the last named, employed for dyeing cotton exclusively, and are readily distinguished from all others and easily estimated, the percentage of ash being an important guide.

To determine the *inorganic mordants*, some 10 to 20 grams of the fabric or yarn are incinerated in a porcelain crucible and the ash calcined for two to three hours until all carbon has disappeared. The method of analysis to be pursued is considerably simplified by the extremely limited number of the mordanting substances of this nature employed; and even the colour of the ash permits a conclusion to be drawn.

Alumina Mordants.—The ash is white. Test by dissolving in hydrochloric acid and adding ammonia. A white precipitate indicates alumina.

Chrome Mordants.—The ash is yellow to brownish-green. Potassium chlorate is added to the ash, which on further heating gives a yellow mass. When this is dissolved in dilute acetic acid and lead acetate is added, a yellow precipitate is formed.

Iron Mordants.—Reddish-brown ash. Dissolve in hydrochloric acid and add potassium ferrocyanide; the precipitate of Berlin blue indicates the presence of iron.

Tin Mordants.—The ash is white, turning yellow on heating. Confirm by heating on charcoal before the blowpipe.

Copper Mordants.—Copper is mostly used in association with iron or chrome. Dissolve the ash in hydrochloric acid, add ammonia in slight excess and filter. When much copper is present the filtrate is blue.

The detection of the mordants can also be effected in the following manner : The sample under examination is immersed in a very dilute solution of bleaching powder until the colour has disappeared. The yellow-brown (ochre) colour of iron, however, remains, and the presence of this metal may be confirmed by the production of Berlin blue on the fibre on addition of yellow prussiate of potash (potassium ferrocyanide). If, after steeping in the bleach solution, the fibres are left perfectly white, then either no mordant, or else only tin or alumina, has been employed. This may be elucidated by dyeing with alizarine. Chrome mordant remains behind after the dye is extracted ; by prolonged exposure to the bleach liquid complete solution occurs.

A scheme for determining the dye stuffs employed is attempted in the annexed tables. If the class of dye stuff present has been ascertained, sample dyeings are performed with the conjectured colours and then treated with the various reagents in current use, the results being compared with the original sample to afford a means of identifying the colours used therein.

The reagents employed are dilute and concentrated sulphuric acid, hydrochloric acid, nitric acid, stannous chloride, ("tin salt") and hydrochloric acid ; dilute caustic soda and ammonia, alcohol and occasionally ether as well.

In many instances the tables on special reactions on the dyed fibre given in the works of Hummel and Knecht, Kertész, and Lehne, are of good service. A knowledge of the colour reactions coming under consideration when a colour has been extracted from the fibre and the further examination of the solution is begun, is also serviceable.

After the dye stuff has been determined, a check experiment by dyeing with the same material should always be performed.

I. BLUE AND VIOLET DYES.

Heat with dilute alcohol and add a few drops of hydrochloric acid.			
No change.	Fibre blue. Liquid blue.		Fibre altered. Liquid yellowish-red.
Indigo blue. Berlin blue. Alizarine blue S.	Indigo carmine and the majority of coal-tar colours.		Wood blue (logwood, fustic). Wood violet (logwood). Orchil violet (orchil). Alizarine violet. Madder violet (madder). Gallëine. Azo-blue.
	New test with concentrated sulphuric acid.		
	Fibre and liquid green.	Fibre and liquid yellow to brown-red.	Fibre unaltered. Liquid blue.
	Methylene blue. Ethylene blue.	Spirit blue. Water blue. Alkali blue. Methyl blue. Victoria blue. Spirit violet. Methyl violet. Acid violet.	Indigo carmine. Induline. Solid blue.

Testing for Pure Indigo Dyes.

1. The water should not be coloured when the sample is boiled therein.

2. On boiling with alcohol, the latter should not become coloured; otherwise aniline colours are present.

3. On boiling with borax or alum solution, no colour is imparted to the solution if indigo alone is present. If the borax solution is coloured red, logwood has been used, and an addition of ferric chloride will give a blue coloration. If the borax solution becomes blue, then indigo carmine or some aniline colour has been used. To distinguish between them add concentrated sulphuric acid. A red coloration indicates aniline, the solution remaining unchanged in the case of indigo carmine.

4. On boiling with weak soda, Berlin blue is recognised by the fibre and liquid becoming more or less brown, a blue coloration being obtained by acidifying the liquid with hydrochloric acid and adding ferric chloride. Indigo carmine is also detected by boiling with soda, the solution becoming deep blue when acidified.

5. On prolonged boiling with glacial acetic acid the indigo is extracted along with all the other colours, and if this extract be shaken up with ether and mixed with water the indigo will be deposited in the lower portion of the ethereal stratum. When indigo alone has been used the subjacent layer of water will be colourless, whereas it will be yellow when dyewood colours have been employed, and violet or blue when aniline colours are present.

II. GREEN DYES.

Heat with dilute alcohol.						
The liquid is colourless:— Indigo and fustic. Wood green (logwood and fustic). Naphthol green. Ceruleine. Solid green.				The liquid is coloured green:— Brilliant green. Malachite green. Methyl green. Alkali green. Light green. Indigo carmine and picric acid. Indigo carmine and quercitron.		
Heat with dilute hydrochloric acid.				Heat with dilute hydrochloric acid.		
Fibre blue. Solution yellow.	Fibre and solution red.	Fibre unchanged. Solution yellow.	Fibre decolorised. Solution yellow.	Fibre paler. Solution blue.	Fibre only slightly changed.	Fibre decolorised. Solution yellow.
Indigo and fustic. Indigo and chrome yellow.	Wood green (logwood and fustic).	Ceruleine.	Naphthol green and solid green.	Indigo carmine and picric acid. Indigo carmine and quercitron.	Light green S. Alkali green.	Brilliant green. Malachite green. Methyl green.

III. RED AND RED-BROWN DYES.

Boil for a short time with dilute alcohol and a fairly strong solution of aluminium sulphate.					
No extract, or else fluorescent extract:— Madder. Eosine. Safranine. Rhodamine.	Yellow to red extract without fluorescence. Add sodium bisulphite:—				
	Immediate decoloration:— Sandal. Redwood. Fuchsine. Fuchsine S. Safflower.		Not decolorised:— Cochineal. Lac dye. Orchil. Anthracene, azo and benzidine dyes.		
	Heat with dilute alcohol.		Heat with very dilute hydrochloric acid.		
	Red extract.	Little or no extract.	Fibre unchanged.	Fibre darkened (brown to blue).	Fibre and solution yellow.
	Sandal. Fuchsine.	Fuchsine S. Redwood. Safflower.	Azo dyes. Orchil. Lac dye. Cochineal.	Congo red. Benzo-purpurin.	Alizarine. Alizarine orange with chrome mordant.
			Heat with dilute lead acetate solution.		
			Fibre unchanged. Orchil. Azo dyes.	Fibre dark brown to violet. Cochineal. Lac dye.	

IV. YELLOW AND ORANGE DYES.

Heat with tin salt (stannous chloride) and hydrochloric acid.			
Fibre unchanged. Solution yellow to colourless.	Fibre decolorised. Solution colourless.	Fibre red to blue-red at first, afterwards decolorised.	Fibre becomes pale yellow. Solution yellow.
Wood dyes and a few aniline colours.	Chrome yellow. Iron buff. Azo dyes.	Fast yellow. Metanil yellow. Orange IV. Brilliant yellow	Alizarine orange.
On heating with lime water.		On heating with ammonium sulphide.		
Fibre reddish or brown.	Little alteration.	Unchanged.	Fibre and solution red.	Fibre blackened.
Fustet wood. Curcuma. Orange (cochineal and quercitron).	Fustic. Quercitron. Flavine. Orleans. Chinoline yellow. Phosphine. Orange from redwood and fustic.	Naphthol yellow S. Auramine. Tartrazine. Citronine. Azoflavine. Orange II. Galloflavine. Chrysoidine. Curcumine S. Chrysophenine	Picric acid.	Chrome yellow. Iron buff.

V. BROWNS, GREYS AND MODE SHADES.

The fibres are heated with dilute oxalic acid and left therein for some time.

The colour is unchanged, solution colourless or slightly coloured (Table A).

The fibre is altered. In dyeing there is always a mordant present (Table B).

Table A.

Combinations of azo dyes, a few aniline dyes, indigo carmine, orchil, etc., differ from the colours given in Table B by

DETERMINATION OF MORDANTS AND DYES. 173

their greater susceptibility to the influence of boiling alkalis (soda or ammonia), whereby partial decoloration and extraction of the dye results.

These combinations are mostly used only for wool and silk; when used for cottons, it is only for very light mode shades.

The sample is warmed with stannous chloride and hydrochloric acid and the change in colour noted.

Fibre and solution colourless or faint yellow.	Fibre blue-red at first, then decolorised.	Fibre and solution fuchsine red.	Fibre blue-grey to blue, restored by washing.
Orange II. Naphthol yellow S. Tartrazine. Orchil. Orchil red A. Bordeaux. Fast red A. Biebr. scarlet. Indigo carmine. Fuchsine. Light green S. (The last two partly recover on washing.)	Orange IV. Fast yellow. Brilliant yellow.	Fuchsine S. Red violet 4 RS.	Induline. Solid blue. Fast blue R. Methyl violet. Acid violet.

The predominant dye stuffs in the fibre give also their characteristic reactions.

The fibre is then further examined for the residual dye, according to the preceding tables, after undergoing a treatment with boiling dilute acetic acid, which brings the colour into greater prominence.

When the fibre is heated with ammonia the following colours are temporarily bleached: fuchsine, fuchsine S, methyl violet, acid violet, red violet, light green S, and also to some extent indigo carmine, whereupon the other dye stuffs become more prominent.

A larger portion of the sample under examination is

boiled repeatedly with dilute soda solution, zinc dust and hydrochloric acid being then added to the liquid, which, after a short time, is filtered.

After filtration the following changes may be observed:—

Colours which, especially on exposure to air, do not reappear.	Colours which recover at once when exposed to the air.	Colours reappearing on neutralisation with sodium acetate and heating the solution.
Azo dyes. Red violet 4 RS. Fuchsine S. Fast blue.	Indigo carmine. Orchil (separated by agitation with ether).	Light green S. Induline. Methyl violet. Acid violet.

If the solution of red violet 4 RS and fuchsine S be reduced a short time the colours reappear on exposure to the air.

If the solution boiled with soda be reduced with zinc dust and ammonia, with corresponding neutralisation, red violet and fuchsine S reappear completely; fast yellow, croceine, Biebr. scarlet, etc., revert to yellow on exposure to the air, and turn red when acidified. The residual coloured solutions can then be examined according to colour reaction tables, or employed to dye fibres which may then be tested further according to the foregoing tables.

To separate the dyed colours a larger sample is boiled repeatedly with soda solution and the extract shaken up frequently and for some time with fresh amyl alcohol. This dissolves out several of the azo dyes, such as orange, Bordeaux, orchil red, etc., whilst indigo carmine, orchil, fast blue, induline, fast yellow, etc., remain in the aqueous solution.

By the aid of this aqueous solution of dye, which is acidified with a few drops of sulphuric acid, some fibre is dyed and the colour tested according to the previous tables. Orchil may be previously separated by shaking up with ether.

The dyes absorbed by amyl alcohol can be recovered by

DETERMINATION OF MORDANTS AND DYES. 175

adding water and evaporating the alcohol, and may then be further examined as above.

Table B.

The sample is heated to boiling with dilute oxalic acid and allowed to remain therein for some time.

Colour almost destroyed, solution nearly colourless.	Colour considerably paler, shade mostly altered, solution yellow to red.	
Iron, copper, alum, etc., lakes from catechu, redwood, fustic, fustet wood, quercitron, logwood, etc.	Chrome, iron, copper, alum, etc., lakes from alizarine colours, catechu, madder, sandal, redwood, fustic, fustet wood, quercitron, logwood, etc.	
Aniline-overdyeing, which is practised especially for brown, grey and mode colours on cottons but less frequently on wool. Alcohol extracts these colours very readily from the fibre, and they can then, when present in sufficient quantity, be detected by Table A and by colour analysis. Any tannin lake present is treated by boiling with soda lye, filtering off the precipitate, and neutralising. (This separates the tannin from the colour.)	The colours of this series differ considerably from those in Table A, in that they are very resistant to the action of boiling alkalis, but on the other hand readily alterable by acids. Any overdyeing, especially of alizarine colours with azodyes, etc., is readily recognised by the extraction of the colour by dilute soda solution or ammonia, which extract can be examined further according to Table A.	
INDIVIDUAL TESTS:— *Tannic acid and catechu.* Boil with yellow prussiate and ammonia: fibre turns brown. *Logwood.* Boil with yellow prussiate and ammonia: partial decoloration. Boil with borax: yellowish-red solution. Add stannous chloride and hydrochloric acid: bright rose-red coloration. *Redwood.* Boil with aluminium sulphate: red extract. Add bisulphite: immediate decoloration. *Fustic.* Boil with aluminium acetate: yellow solution with green fluorescence.	Boil for a long time with a solution of yellow prussiate and ammonia.	
	Fibres very slightly changed.	Fibres paler.
	Alizarine. Alizarine S. Alizarine orange. Alizarine chestnut. Anthracene brown. Galleine. Galloflavine. Madder, Sandal. Catechu, Fustic. Quercitron.	Alizarine blue. Ceruleine. Alizarine black. Redwood. Logwood. (Much altered, paler.) *N.B.* If the sample, as is the case with many wood dyes, has been darkened with iron, the fibres will always become paler.

A larger portion of the sample to be tested is repeatedly boiled for a long time with dilute hydrochloric acid and finally with dilute soda solution. The following shades of colour are very resistant towards dilute acids : Alizarine blue, alizarine black, anthracene green, sandal, and madder (to some extent). The larger proportion of the dye stuff remains on the fibre, and can be identified according to the corresponding tables.

Both extracts are next neutralised with soda lye and boiled, whereupon the following phenomena appear :—

The colour lakes dissolved in hydrochloric acid and soda separate immediately and almost completely. The filtrate is slightly coloured.	The colour lakes separate out only to a slight extent at once. The filtrate is more or less strongly coloured.
Alizarine chestnut (bluish-red p.p.). Anthracene brown (dark grey p.p.). Ceruleine (green p.p.). Alizarine blue (greenish-blue p.p.). Alizarine black (blue-grey p.p.). Galleine (violet p.p.). Catechu, fustic, quercitron.	Alizarine (red solution). Alizarine orange (red solution). Madder (red solution). Galloflavine (yellow solution). Sandal (red solution, minute p.p.). Redwood (yellowish-red solution). Logwood (brownish-red solution).
N.B. The precipitates may be freed almost entirely from logwood deposit by careful treatment with a warm solution of yellow prussiate.	Alizarine and alizarine orange deposit from the above solution, after twelve hours' standing, in flakes, especially when the liquid is strongly acidified. The precipitate is sublimed and further identified. (Soda solution gives a violet coloration with alizarine.)

INDIVIDUAL TESTS : *A*. ANTHRACENE COLOURS.

The well-washed precipitate is treated with concentrated sulphuric acid for some time, the solution being then considerably diluted and left for six or eight hours. The colour lake is decomposed, the colour separating out in flakes on standing, which when filtered, washed, and (when in sufficient amount) dried and sublimed, are further identified by analysis. Or the precipitate may be dissolved at once by a few drops of soda lye and the colour tested further, whereby, however, regard must be had to any impurities present. The precipitate is coloured on the filter as follows, on addition of concentrated soda lye : alizarine, violet; chestnut and orange, bluish-red.

The determination can also be effected in the following manner :—

A larger amount of the sample under examination is boiled repeatedly with dilute sulphuric acid and then with dilute soda solution, the extracts being united and evaporated to dryness. The residue is treated for some time with sulphuric acid, greatly diluted, and after six to eight hours' standing the precipitate is filtered off, washed, dried, sublimed and the dye tested.

INDIVIDUAL TESTS : B. WOOD DYE STUFFS.

Madder.—Boiling with aluminium sulphate gives a solution with yellow-green fluorescence, disappearing on addition of glacial acetic acid.

Catechu (on cotton only).—In presence of chrome mordants a characteristic resistance to dilute acids, especially organic acids, is manifested.

Sandal.—Aluminium sulphate gives a red extract decolorised by sulphurous acid. Alcohol gives a red extract.

Redwood.—Aluminium sulphate gives a red extract, decolorised by sulphurous acid. Alcohol gives no extract. Readily abstracted by dilute acids and alkalis.

Logwood.—The fibre becomes much paler on boiling with yellow prussiate solution and ammonia. A reddish-yellow solution is obtained by borax solution ; neutralised with acetic acid and treated with ferrous sulphate solution, a black coloration or precipitate is gradually formed. Boiling with dilute acetic acid gives a yellowish-red solution, which, on cooling and addition of stannous chloride and hydrochloric acid, yields a rose-red coloration. The resistance to dilute acids is slight.

Fustic.—On boiling with aluminium acetate, a yellow solution with strong green fluorescence is produced, and is not destroyed by sulphurous acid.

VI. BLACK DYES.

Heat with dilute hydrochloric acid.	
Fibre and solution red to yellow.	Fibre unchanged.
Logwood black. Logwood black with indigo ground. Tannin black. Catechu black. Alizarine black. Resorcine black.	Alizarine black (Naphthazarine), solution greenish-blue. Aniline black. Wool black, solution pale blue. Naphthol black, solution reddish. Cachou de Laval, solution faint grey. Brilliant black, solution faint reddish-violet.

XII. DETECTION AND ESTIMATION OF ARSENIC.

Some 25 grams weight of the sample of fabric or yarn under examination is taken and placed in a tubulated 500 *c.c.* retort connected with a condenser. The latter debouches through a cork in the neck of a 600 *c.c.* flask, which contains 100 *c.c.* of water, and is cooled by immersion in cold water. The flask is also connected with a series of bulb tubes, also containing a little water. At the outset 200 *c.c.* of pure hydrochloric acid are poured over the substance in the retort, and at the end of an hour 5 *c.c.* of a concentrated solution of ferrous chloride are added, the retort being then heated in an oil bath until the contents have distilled over, with the exception of a small residue. When cold, another 50 *c.c.* of hydrochloric acid are added and the distillation repeated. One portion of the distillate is examined in Marsh's apparatus.

From 20 to 25 grams of zinc rod are immersed for a few minutes in platinum chloride solution and well rinsed with water, after which they are placed in a 250 *c.c.* flask into the neck of which a tube is fused at right angles for connection with a calcium chloride tube and a piece of very refractory glass tubing, about 30 *cm.* long and 6 *mm.* wide, enveloped in wire gauze, the tube being drawn out to a point at the end and bent upwards. A drop funnel in the neck of the flask supplies dilute hydrochloric acid drop by drop to the zinc, and after the evolved hydrogen has streamed through the apparatus for a short time to drive out the air, the refractory tube is heated to redness, between the 15th and 20th *cm.* from the end, by four or five Bunsen burners placed side by side, and the gas escaping from the pointed end is ignited.

A measured quantity (about 50 or 100 *c.c.*) of the above-mentioned distillate suspected to contain arsenic is placed in the drop funnel and run into the flask at the rate of 1 *c.c.* per minute. The arseniuretted hydrogen formed is decomposed in the heated tube, and a deposit of metallic arsenic settles in the colder or drawn-out portion of the tube. When all the distillate is run in, the funnel is rinsed with 10 to 20 *c.c.* of hydrochloric acid; the tube is then allowed to cool and the part containing the mirror of metallic arsenic is cut off, weighed, and re-weighed after the arsenic has been driven off by heat, the difference giving the weight of arsenic in the portion of distillate. From this the amount in the entire distillate or in the sample can be easily calculated.

Before applying the arsenic test, a blank experiment must be performed in order to ascertain whether the reagents are pure or contaminated with arsenic.

APPENDIX.

OFFICIAL SPECIFICATIONS FOR THE SUPPLY OF MATERIALS FOR USE IN THE GERMAN ARMY.

1. Cloth.

Constitution of the Cloth.—Only good, sound fleece wool is permissible, the admixture of shoddy, wool from dead sheep, or tanner's wool being prohibited. The cloth must be of good appearance, durable, as free as possible from faults, and correspond with the sample in thickness of material, quality of wool, colour, mingling and thorough dyeing of the yarn, nature and closeness of texture of the fabric, fulling, washing, raising and shearing, on both right and wrong sides.

The mixed grey cloth, the dark green kersey and the dark blue mixture, plain or twilled, must therefore be delivered in an unroughened state, but milled on both sides, whereas all the other kinds of cloth must be milled on the under side only, being raised and cropped on the face.

The sample of dark blue cloth No. 1 prescribes the quality of the wool to be used for Moltons.

The white kersey and white cloth must not be sulphured, and the latter must not be chalked to a greater degree than the sample.

The samples supplied by the Clothing Department are only intended to show the kind of cloth and shade of colour required.

The normal average weights of the cloths are :—

Dark blue mixture, plain or twilled cloths,
                             715 grams per metre (= 23 oz. per running yd.).
Grey mixture cloth -   -   638   ,,   ,,   ,,   (= 20·6 oz. per yd.).
Dark blue cloth No. 1, cornflower blue, Russian blue, brown, dark green, ponceau red cloth No. 1, madder red cloth No. 1 and black cloth No. 1 -   -   -   613   ,,   ,,   ,,   (= 19·8 ,,   ,,).
Dark blue cloth No. 2   -   588   ,,   ,,   ,,   (= 19   ,,   ,,).
All grades of badge cloth   525   ,,   ,,   ,,   (= 17   ,,   ,,).
White kerseys   -   -   750   ,,   ,,   ,,   (= 26·1 ,,   ,,).
Dark green kerseys -   -   624   ,,   ,,   ,,   (= 20·1 ,,   ,,).
Dark blue Moltons   -   -   550   ,,   ,,   ,,   (= 18   ,,   ,,).

When the cloth exhibits special points as regards quality of wool, evenness of threads, good shotting, small selvedges and fine colour, the following margins of underweight are allowed :—

Dark blue mixture -   -   19 grams per metre (= 0·61 oz. per yd.).
Grey mixture, dark blue Nos. 1 and 2 (Molton excluded), cornflower blue, Russian blue, brown, dark green, ponceau red No. 1, madder red No. 1 and black No. 1   25   ,,   ,,   ,,   (= 0·8 ,,   ,,).
Badge cloths   -   -   -   12   ,,   ,,   ,,   (= 0·39 ,,   ,,).

The width of all cloths, except the dark blue mixture, plain and twilled cloths and dark blue Moltons, shall measure 1·17 m. (= 46 in.) exclusive of the selvedges, that of the dark blue mixture, plain and twilled cloth and Molton being 1·28 m. (= 53½ in.) between selvedges.

In length the pieces must measure from 20 m. (21⅞ yds.) minimum to 27 m. (29½ yds.) maximum ; Moltons alone being exceptionally admitted in longer lengths, when of unimpeachable quality. The decimetre (3·9 in.) is the smallest unit of length used, so that centimetres, 1 to 9, are disregarded in measuring a piece.

Ponceau red cloth No. 1, as well as the coloured badge

cloths, must be delivered shrunk, all the other kinds unshrunken. The latter must not lose more than 5·5 cm. per metre (i.e., 5½ per cent.) in the length and 4·2 cm. (1·65 in.) in the width of the cloth during shrinking.

Warp.—The warp threads in the prescribed widths must total, for :—

Grey mixture cloth	1700 threads.
All foundation cloths for tunics, etc., including white kerseys	2000 ,,
Dark blue mixture cloth, dark green kersey, dark blue Molton and badge cloths	2200 ,,

Weft.—A thread of lower twist than for warp is permissible, but the yarn for shotting must be quite as evenly spun.

Ponceau red cloth No. 1 must be cold pressed, but all the other cloths may be pressed moderately warm; steaming the cloth is prohibited.

The *colours* must be thoroughly *fast*, *i.e.*, must be able to withstand the everyday influence of light, air and water for a suitable time.

Whilst the mixed cloths, the dark green kersey, the light blue cloth and the dark blue Molton must be dyed *in the wool*, the other kinds are only required to be *piece dyed;* these should, however, in so far as the shade of colour desired permits, be commenced in the wool, in order that the requisite penetration of the dye may be secured. The blueing of the dark blue cloths Nos. 1 and 2, the cornflower blue, the dark green cloths and the dark blue piece-dyed Moltons must be effected with *pure indigo*, as also the finishing dyeing of the dark blue stuffs, the employment of any addition of coaltar colours (aniline) for the blue cloths being prohibited.

To show that the cloths have been sufficiently blued in the wool, ten to fifteen white threads must be woven into the front end of each piece.

The piece-dyed cloths must be evenly dyed and may not

be streaky. The brown cloths must not show up pale in section, but must be dyed through.

All pieces must be thoroughly rinsed, and must not lose colour more than the sample. The dark green kerseys must not lose colour either in the wet or dry state.

The minimum tensile strength and elasticity (warp and weft) of the various cloths are given below, the sample (doubled) to be 9 *cm.* (3·53 in.) wide and 30 *cm.* (12 in.) long between the clamps of the dynamometer.

Tensile Strength.

		Weight.
White kersey - - - - - - - -	70 kilos.	(= 154 lbs.).
Foundation cloth for tunics, etc., and grey mixture cloth - - - - - - - -	56 ,,	(= 123¼ ,,).
Dark blue mixture cloth and dark green kerseys -	60 ,,	(= 132 ,,).
All badge cloths - - - - - - -	46 ,,	(= 101¼ ,,).
Dark blue Molton - - - - - - -	42 ,,	(= 92·4 ,,).

In cases of special excellence in point of quality of wool and careful manufacture, a margin of 3 kilos. (6·6 lbs.) will be allowed off the foregoing figures, except for Moltons, which must always have the above minimum strength.

Elasticity.

For kerseys and dark blue mixture cloths - - - -	10 *cm.* (4 in.).
,, dark blue Moltons - - - - - - - -	7 ,, (2¾ ,,).
,, all other foundation cloths and all badge cloths - -	6 *cm.* (2·36 ,,).

The above figures represent the minimum permissible limits.

The *selvedges* must not be of calves' hair, and must agree in weight and colour with those of the standard samples.

All foundation cloths must have the name of the firm and the manufacturing number sewn in before fulling.

Regulations to be Observed on Taking Delivery.

1. Notice should be taken that deficiencies in weight have not been made up by damping the cloth or by broad, thick selvedges.

2. The width of the cloth should be measured by folding the piece and taking the half-breadth.

3. On reception, a sample of 1 to 3 metres in the full or half-breadth of the cloth from each parcel, particularly the foundation cloths, should be shrunk, the process adopted being left to the discretion of the clothing officials.

The loss in length sustained by the cloth during shrinking must be charged to the contractor. The shrinking must be tested by the Clothing Department officials.

4. In testing the permanence of the dyes, the following conditions must be observed :—

A cutting of the cloth is kept for two minutes in pure dilute (1 : 3) hydrochloric acid until thoroughly impregnated and then at once rinsed in cold water. The colour of the dark blue cloths Nos. 1 and 2, the pale blue, Russian blue and potash blue cloths and dark blue Moltons should be unchanged, whereas the brown colours may become pale brown and the greens bluish-green, and the brilliancy of the cornflower blue colour and the depth of the black in dark blue mixture and grey mixture cloths may be diminished. The colour of the black cloths must become reddish, whilst the crimsons may assume the colour of an impure ponceau red, and the pompadour, ponceau and rose-reds, as well as the colours of all madder red and yellow cloths, may become somewhat fainter. In addition,

The Black Cloths

are to be tested with alum, 10 grams of this substance being dissolved by boiling in a three-fold quantity of water and, after cooling down to 50° R. (62·5° C.), employed for steeping a sample of the cloth for two minutes until the latter is thoroughly impregnated, whereupon it is at once rinsed with cold water. The colour should not undergo any appreciable change under this treatment.

The Ponceau Red Cloths

are tested with ammonia in the manner laid down for the dilute hydrochloric test. The colour can only be regarded as fast and acceptable provided it assumes a ponceau red or violet tinge.

The Yellow Cloths

are tested by immersing a sample strip 20×5 cm. (8×2 in.) in a glass half full of soft cold water for an hour, whereby the water should not become tinged with colour to any extent; should, however, the water be stained greenish-yellow or yellow after five or ten minutes, the cloth must be rejected.

The Light Green Cloths

are tested with water in the same way, except that the immersion extends for two hours. If the water does not become coloured, or if a piece of white cloth enveloping the green sample is not stained, the colour must be considered as sufficiently fast.

5. Out of each delivery and each kind of cloth composing same not less than 10 per cent. of the pieces (minimum, four pieces) must be tested in the dynamometer. If more than one-fourth of the samples fail to pass the test for weft strength and elasticity, then each piece in the parcel must be tested. In the case of the warp tests, when deficiency in quality is recorded, the contractor shall be asked whether he is willing to stand the loss resulting from further testing the warp or prefers to take back the goods.

6. When spots or nops are found to have been disguised with nop tincture, or nops and other small holes are found to have been filled up, or when deficient weight has been supplemented artificially, these practices shall be considered as deliberate attempts at deception.

2. Linens and Cottons for Military Use.

(a) *Linens.*

The yarns employed for the linen cloths must consist of pure flax fibre, unmixed with any extraneous material, and free from any woody particles and shives.

By "flax yarns" shall be understood those prepared from the long hackled fibres of flax regularly and solidly spun, and known in the trade as "line". "Tow yarns" are those prepared from the residual short fibres, and known commercially by this designation.

The yarns entitled "three-quarter bleached" must be of a good white colour and free from woody particles.

Bleached, creamed and dyed linen warps from short fibre must not be more heavily sized than is absolutely necessary. (Long-staple warps and those for unbleached goods, as well as wefts of all kinds, must not be sized at all.)

The fabric must be well and evenly woven, with unbroken selvedges. In hand-woven goods small defects in appearance or irregularities form no ground for rejection.

Only the yarn numbers given below may be used for making the goods. The use of other numbers or descriptions of yarns shall be regarded as an intentional deception, and when proved shall result in the exclusion of the contractor from tendering in future.

Apart from the prescribed mangling and the necessary clearing of shives, etc., unbleached linens must not exhibit any finish nor contain more than the natural quantity of moisture.

Bleached or re-bleached and dyed goods must not be loaded in the piece with more starch or other dressing than is contained in the sample.

Calendering (smoothing the cloth in single layers between rollers), instead of mangling, is not allowed.

Blue linen linings blued with indigo to light blue in the piece must not, when rubbed with white paper, part with more colour than the sample. The dye must exhibit a sufficient degree of fastness when treated with warm dilute (1 : 10) hydrochloric acid.

Each piece must have as a termination at either end some 20 weft threads differing in colour from the rest of the wefts.

DESCRIPTIVE LIST.

Stuff.	Width in cm. (of 0·39 in.).	Kind of Yarn.	Yarn No.	No. of threads per 1 sq. cm. (0·154 sq. in.).	Weight per running metre in grams (= oz. per yd. × 31).	Finish.	
						Whether unbleached, bleached, or dyed.	Whether and how far mangled.
Drills for jackets	83/84	Unbleached yellow flax	20	20-21	340-350	Unbleached	Medium mangling
		Unbleached grey tow	14	18-19			
Drills (better class)	83/84	Unbleached yellow flax	25	26-27	290-300	Unbleached	Medium mangling
		Unbleached grey tow	25	22-23			
Drill for trousering (diagonal)	75/76	Unbleached yellow flax	25	23-24	295-305	Unbleached	Medium mangling
		Unbleached grey tow	16	19-20			
White linen for trousers	75/76	¾ bleached flax	25	20¼-21¼	180-190	Rebleached	Heavily mangled
		¾ bleached flax	28	19¼-20¼			
Canvas for trousers	75/76	Unbleached yellow flax	16	15¼-16	295-305	Unbleached	Lightly mangled
		Unbleached yellow tow	14	16-17			
Grey linen linings	75/76	Unbleached grey tow	18	14¼-15	210-220	Unbleached	Heavily mangled
		Unbleached grey tow	20	16-17			
Blue linen linings	77/78	Unbleached grey flax	28	16¼-17	185-195	Dark blue dyed in the fibre with fast indigo	Heavily mangled
		Unbleached grey tow	30	17-18			

(*b*) *Cottons.*

The yarns used for cottons must be evenly spun with good twist from good, clean long-staple cotton, and should contain no woody particles and only a minimum of shives.

Warp yarns may only be sized to a degree absolutely necessary for manipulation; weft yarns not at all.

Cotton fabrics, with the single exception of dyed calico linings, may not exhibit any finish in the piece apart from the necessary mangling and cleaning the surface from particles of seed and weaving flocks.

The finish of the blue calico linings must correspond exactly with that of the standard sample, or, when required unstiffened, with the conditions of the order.

The degree of mangling required is indicated by the sample or in the subjoined list. Calendering, in place of mangling, is prohibited.

The goods must be evenly and well woven and the selvedges entire. Small irregularities and defects in appearance in the case of hand-loom goods shall not constitute a ground for their rejection.

Only the yarn numbers given in the appended descriptive list may be used, the employment of any others being regarded as an attempt at fraud, and will result in the disqualification of the contractor from tendering in future.

The requirements as regards dyeing of blue calico linings are identical with those laid down for dyed linens. Each piece must also have a number of differently coloured weft threads at each end as prescribed for the linens.

DESCRIPTIVE LIST.

Stuff.	Width in cm. (of 0·30 in.).	Kind of Yarn.	Yarn No.	No. of threads per 1 sq. cm. (0·154 sq. in.).	Weight per running metre in grams (= oz. per yd. × 31).	Finish.	
						Whether un-bleached, bleached, or dyed.	Whether mangled and to what degree.
Calico shirting	}75/76	Cotton	16-17 / 11-12	23-25 / 23-25	170-180	Un-bleached	Un-mangled
D.W. calico for drawers	}75/76	Cotton	18-19 / 14-15	21-23 / 17-19	190-200	Un-bleached	Lightly mangled
D.W. twill for drawers	}75/76	Cotton	18-19 / 16-17	18-20 / 24-26	180-190	Un-bleached	Lightly mangled
Blue calico linings	}77/78	Cotton	16-17 / 18-19	22-24 / 22-24	170-180	Dark blue (fast dyed with indigo in the fibre)	Heavily mangled
White calico linings	}75/76	Cotton	16-17 / 18-19	24-26 / 24-26	145-155	Un-bleached	Heavily mangled
D.W. white calico linings	}75/76	Cotton	18-19 / 18-19	21-23 / 21-23	190-200	Un-bleached	Heavily mangled

Remarks.

In the preceding tables the figures referring to the yarn number and count of the cloth are separated by horizontal lines, those above the lines being for warp, and those below, weft, threads.

The yarn numbers given are those of the English system of numbering.

The lengths of the pieces shall be those current in the trade as "full length pieces" of the various goods.

A margin of ± 3 per cent. is allowed on the weight provided the stuff is in other respects equal to the standard sample.

When the calico shirtings are ordered wider than the breadth specified, the weight must increase by 2·3 grams per metre for each additional 1 *cm.* in the width (0·19 oz. per yd. for each additional 1 in.).

Regulations to be Observed in Taking Delivery.

1. Each piece must be marked by the official with the factory number and also with a consecutive number if necessary.

On delivery, before the goods are taken in, they must be examined for damage from infiltrations of water, fat, etc., or from the use of iron hooks. If such be discovered, the goods must only be accepted under reserve, in order to protect the interests of the contractor and fix the liability of the railway company or the other carrier.

2. The fibres employed for spinning must be examined by analysing several threads and subjecting them to microscopic examination, in order to determine their nature (whether flax, cotton or other fibre) and quality (length of staple, durability, etc.).

Flax yarn may be readily distinguished by its regularity and lustre from the uneven, rougher and always rather knotty tow yarn, either by holding the fabric up to the light or taking out a few threads.

3. It is specially important to determine whether the provisions of the specification have been complied with as regards the sizing of the yarn and the dressing of the fabric. In this respect guidance is afforded by the estimation of the dry weight of a sample cutting before and after four or five thorough boilings, washings and rinsings. (In the case of cottons, the loss in weight should, as a rule, not exceed 10 per cent.)

In continuing the examination, the presence of added mineral matter is detected by weighing the ash, whilst vegetable substances are determined by chemical tests.

If the results obtained cause prohibited weighting to be

suspected, the opinion of an expert must be taken, in view of the consequences attaching to the practice of intentional deception on the part of the contractor, before the whole delivery is rejected, and a notification of the circumstances drawn up.

4. The number of threads is determined by the aid of the thread counter, but the warp threads should not be counted too near the edge nor the wefts too near the ends of the cloth.

5. The finish of the stuff is examined by comparison with the sample. It should be borne in mind that heavy mangling makes the cloth firmer, closer and more opaque, whilst light mangling leaves it more open; so that in the case of heavily mangled stuffs, good feel is no proof of good quality.

6. The colour testing to be performed on the dark blue linings results from the requirements imposed as regards the thorough dyeing (see p. 188) of same.

7. Concerning the extent of the tests to be performed, the following remarks apply :—

At least one-tenth part (four pieces minimum) of each delivery shall be examined. It is not necessary that each piece should be examined for all the properties required by the contract, but different pieces may be tested for different properties.

3. WATERPROOF MATERIALS FOR MILITARY PURPOSES.

(a) *Canvas for Laced Shoes.*
Description.

The canvas is a close unbleached fabric required to possess the following properties :—

Kind of Yarn.—For the warp a double-twisted No. 10 flax yarn, dry spun from good long flax, must be used, and be free from more than a minimum of knots, shives or felted places; for the warps a similar quality No. 12 yarn is required, but in this case a best dry-spun single long-staple hemp yarn may be used instead of flax.

If the canvas is found to contain—even a single piece—any other fibres, especially cotton or jute, or inferior material such as tow, or if wet-spun yarn has been used, the entire parcel will be rejected.

The count of threads per square centimetre shall be $12\frac{1}{2}$-13 for warps and $9\frac{1}{2}$-10 for wefts, and is to be ascertained by the aid of the thread counter.

Weaving.—The warp and weft threads must appear in regular alternation on the face of the cloth. The fabric must not be irregular or defective, and must be woven on heavy, so-called power looms.

Breadth, Length and Weight.—The canvas must be 75 cm. ($29\frac{1}{2}$ in.) wide and in pieces of the length customary in the trade, the weight being about 650 grams per running metre (21 oz. per yd.) with a margin of $\pm$ 3 per cent. Each piece must terminate with a border of about 20 weft threads at each end.

Finish.—Light mangled finish is required, but calendering, even with wound rollers, is prohibited.

Sizing the yarn, as well as the use of loading materials (such as starch, etc.), will be regarded as fraudulent, and dealt with as laid down in the conditions of the tender.

Breaking Strain, Elasticity.—Strips of the fabric 5 cm. (2 in.) wide and 36 cm. (14 in.) free length taken from any convenient part of the stuff must stand a strain of at least—

and
    220 kilos. (484 lbs.) in the direction of the warp
    260 ,, (572 ,,) ,, ,, ,, weft.

A minimum of elasticity is to be regarded as preferable.

The fabric should be sufficiently permeable to gases, but, on the other hand, must not permit the transfusion of water beyond the following limits:—

A square piece of cloth of 25 *cm.* (10 in.) side, cut from any part of the piece, folded like a paper filter, placed in a 60° glass funnel and loaded with 300 *c.c.* of distilled water, must not at the end of twenty-four hours show on the under side more than very small evenly-distributed drops of water without the stuff itself being wet through.

The means employed for rendering the stuff watertight are left to the choice of the contractor, with the proviso that the use of substances dangerous to health is prohibited.

Remarks.

The testing of the weight and breaking strain shall be performed in an apartment where the atmospheric moisture averages 50 to 70 per cent. (measured by the Koppe-Saussure hygrometer) and after the goods have been exposed therein loose or unrolled during at least two hours.

In all cases of dispute over the constitution of the goods the decision of the Royal Mechanical Experimental Institute at Charlottenburg shall be taken as final. The samples to be examined shall be sent by the clothing officials to the Institute, and the costs of the carriage and testing shall follow the result.

Regulations to be Observed on Taking Delivery.

1. The pieces are numbered on reception and examined for damage from water, fat, etc., as prescribed in the case of linens and cottons.

2. The nature of the yarn must be tested as follows :—

(*a*) A few threads are first drawn from the fabric to ascertain whether dry- or wet-spun yarn has been used. Dry-spun material is recognised by the soft and fibrous appearance of the yarn, the cloth being also more supple and closer in texture. The wet-spun thread is hard and smooth, not woolly, and the cloth is also stiff and hard, and when held up to the light displays many open spots.

(*b*) Next follows the examination of the fibre. Whether long-staple flax (hemp) has been used can be easily determined by unravelling and dividing individual threads.

(*c*) The presence of any other fibre than flax (or hemp) can be detected by the microscope. To perform this examination thoroughly the following points have to be observed:—

Preparing the Material.—About five warp and weft threads 10 *cm.* (4 in.) long are boiled $\frac{1}{4}$ hour in a $\frac{1}{2}$ per cent. solution of caustic soda and then cleaned by washing in water.

Preparing the Object.—A small portion is scraped with a knife from each of the warp and weft threads and placed on a glass slide in one or two drops of potassium iodide-iodine solution (20 grams water, 1·15 grams iodine, 2 grams KI, 2 grams glycerine), wherein it is spread out by dissecting needles and then covered with a cover-glass. After soaking up the excess of iodine solution round the edges of the latter, the object is examined under a power of 300.

It is advisable to have a number of scrapings of flax, jute, cotton, etc., in bottles of distilled water at hand for comparison.

An adulteration with jute can also be readily detected by immersing a number of warp and weft threads in phloroglucin solution (3 grams phloroglucin, 25 *c.c.* alcohol, 25 *c.c.* concentrated hydrochloric acid). At the end of $\frac{1}{4}$ hour jute threads will be coloured a dark red, whereas hemp or flax will be coloured merely a faint rose at most. The coloration disappears after prolonged standing.

3. Whether the stuff has been mangled or calendered can be recognised, in that by the latter process, wherein the cloth runs singly between rollers, the threads are flattened out, and, even if the rollers were not wound, receive a certain amount of gloss.

4. To determine whether the goods have been weighted with starch, cuttings, 1 square decimetre (3·93 sq. in.) each, are taken from three different places in the piece, and boiled for ¼ hour with 1 litre of distilled water, after which five drops of iodine solution (20 grams water, 1 gram iodine, 2 grams potassium iodide) are added to the liquid. If the latter turns blue, the presence of starch is proved.

Other loading materials are seldom used, since they retard impregnation; but if suspected as present, must be tested for by a chemist.

5. The strips to be tested in the dynamometer are cut a little wider than the prescribed width and reduced to the exact size by removing a few threads from either side. A minimum length of 60 $cm.$ (23½ in.) should be cut and exactly 36 $cm.$ (14 in.) marked off by ruling two lead pencil lines across the strip, leaving 12 $cm.$ (4¾ in.) at each end for fastening to the machine. In getting ready for the test, care must be taken that the clamps exactly coincide with the pencil markings when the tension is applied. Subjecting some of the threads to a greater tension than the others, as a result of irregular adjustment of the sample, must be avoided, it being essential that the tension should be evenly distributed over the entire breadth of the strip. If the strip should break close against either of the clamps the test must be repeated with a fresh sample; the fracture should take place at least 1 $cm.$ (0·39 in.) away from the clamp. Moreover, the result should never be taken from one test by itself.

6. One-tenth portion of each delivery shall be tested, but it is unnecessary to submit each piece to all the various tests required.

(*b*) *Stuffs for Provision Bags and Knapsacks.*

Description.

Brown cotton No. 10 doubled yarn is to be used for warp

and weft, and both must be evenly spun from good long-staple cotton.

If even a single piece of any other fibre than cotton is detected, the entire parcel will be rejected.

The *counts of thread* per sq. cm. shall be : warp, 24 to $25\frac{1}{2}$; weft, $12\frac{1}{2}$ to $13\frac{1}{2}$.

Weaving.—The warps must be doubled, so that looking along the direction of the weft each warp appears as a pair of threads alternating on the face with the weft. The fabric must be evenly woven and free from faults.

Provision-bag cloth must be 92-93 cm. (36 in.) wide, the pieces being of the usual lengths current in the trade and the weight about 470 grams per running metre (15 oz. per yd.), the margin of allowance being in this respect ± 3 per cent.

Each piece must be bordered at both ends with some twenty undyed weft threads, and is required to be mangled, calendering not being permissible.

The minimum breaking strain (for samples of the usual dimensions) is 115 kilos. (253 lbs.) for the warp and 75 kilos. (165 lbs.) for the weft.

The degree of permeability by water must not exceed the limits prescribed for canvas for laced shoes.

The dye employed must be fast catechu, without any addition likely to be injurious to health, and must be applied to the reeled yarn before weaving. If it appears that the goods have been dyed in the piece, they must be rejected, and no pieces which have been unevenly dyed or which differ appreciably from the standard sample can be accepted.

Regulations to be Observed on Reception.

1. The pieces are to be numbered and examined for damage, as already mentioned under linens, etc.

2. Testing the purity of the cotton is performed micro-

scopically, but as cotton is one of the cheapest of the textile fibres, and therefore little likely to be adulterated, it will be sufficient to test only two pieces. On the other hand, the length of staple of the cotton employed must be examined within the usual limits.

3. Finish, loading and breaking strain are examined in the usual manner, as above.

4. The quality of the catechu dye is tested in the following manner :—

(a)[1] One square decimetre of stuff, cut into pieces if necessary, is boiled in $\frac{1}{2}$ litre of distilled water for five minutes, whereby the water should be at most merely slightly tinged with brown. When the same sample is boiled with a fresh quantity of water the latter should not assume any coloration whatever.

On drying, the sample should exhibit unaltered its original colour.

(b) The same results should be obtained when the stuff is boiled with 80 per cent. alcohol on the water bath.

(c) A smaller strip of the cloth (1×3 cm.) is boiled along with acetic ether in a test tube by immersion in hot water—the flame having been extinguished on account of the inflammability of the ether. The solvent should not be coloured.

(d) One square decimetre, cut into strips if necessary, is boiled for five minutes in 100 $c.c.$ of an 8 per cent. solution of alum, whereby the latter may become somewhat brown in colour, but the stuff when washed in clean water and dried should only appear a little lighter. One-half of the sample treated with alum is steeped for five minutes in a 5 per cent. caustic soda solution and when washed and dried should then appear appreciably darker than the other portion,

---

[1] To ensure the quick and regular working of the aqueous reagents (water, alum and oxalic acid), the sample should have just previously been steeped in alcohol to drive out the interstitial air, and afterwards rinsed with water.

although not quite so dark as the original colour. Other colours, such as anthracene brown, also darken under the action of caustic soda, but the resulting shades vary to such an extent from the original that there is no danger of confounding these with catechu.

(e) The stuff should not become appreciably lighter when treated for $\frac{1}{4}$ hour with concentrated oxalic acid at the ordinary temperature.

The goods are only considered as acceptable when the requirements of the contract are conformed to by the results of every test employed.

When the colour of the pieces is regular only one of them need be tested for the quality of the catechu, or each test may be applied to a separate piece.

The distinction between yarn-dyed and piece-dyed goods can be recognised by examining the cut ends of the threads, the centre of which will be paler in the latter event. The regularity of the colour of the stuff is tested by examination whilst it is passing over the roller.

(c) *Tent Canvas (for Portable Tents).*

Doubled best quality No. 20 cotton twist, evenly and well spun, must be used for warp and weft.

If any other fibre than cotton is found, even in a single piece, the entire parcel will be rejected.

The count per sq. *cm.* must be 20-21 threads in warp and weft.

Weaving.—One warp and one weft must be alternately visible on the face looking along the weft thread.

Width of tent canvas, 94-95 *cm.* (37 in.), and the pieces of ordinary commercial length ; weight, 265 grams per running metre (8½ oz. per yd.) with a margin of ± 3 per cent.

Each piece must terminate at either end with a border of

some twenty undyed weft threads, and the goods must be of light mangled finish.

The breaking strain (samples of usual size) must be at least 60 kilos. (132 lbs.) for the warp and 65 kilos. (143 lbs.) for the weft.

Permeability to water and dyeing same as for provision bagging, and the same regulations for taking delivery apply.

4. Linens and Cottons for Barrack and Hospital Use and for the Training Colleges.

Description.

The observations already made on p. 187 apply equally in the present instance. Cottons for neckcloths may have a moderate amount of dressing.

A. LINENS.

Stuff.	Width in *cm.*	Kind of Yarn.	Yarn No.	No. of threads per 1 sq. *cm.*	Weight per running metre in grams.	Finish. Whether unbleached, creamed, yarn-white, rebleached or dyed.	Finish. Whether and to what extent mangled.
Grey linen for bed palliasses and mattresses (for barrack and hospital use)	100 or 104	Unbleached grey tow / Unbleached grey tow	12 / 10	11 / 12½	395-405 or 410-420	Unbleached	Unmangled
Grey linen for bolster cases (barrack use) and pillow covers (hospital use)	83	Unbleached grey tow / Unbleached grey tow	12 / 10	11 / 12½	325-335	Unbleached	Unmangled
Grey linen for lining ordinary invalids' coats and trousers	70	Unbleached grey tow / Unbleached grey tow	20 / 22	14 / 15½-16	170-180	Unbleached	Unmangled
Grey linen for aprons	100	Unbleached grey tow / Unbleached grey tow	16 / 16	14 / 15-16	335-345	Unbleached	Unmangled
White linen for ordinary sheets	68	¾ bleached tow / ¾ bleached tow	18 / 20	16 / 16	170-180	Yarn-white	Unmangled
White linen for ordinary covers and bolster cases	84	¾ bleached tow / ¾ bleached tow	20 / 20	17 / 17	215-225	Yarn-white	Unmangled
White linen for fine sheets	68	¾ bleached flax / ¾ bleached flax	25 / 28	21-22 / 20-21	170-180	Rebleached	Medium mangling
White linen for fine covers and pillow cases	84	¾ bleached flax / ¾ bleached flax	25 / 28	21-22 / 20-21	210-220	Rebleached	Medium mangling
Stuff for fine towels	42	¾ bleached tow / ¾ bleached tow	20 / 18	21 / 21-22	130-140	Rebleached	Unmangled
Stuff for ordinary towels	52	Creamed tow / Creamed tow	12 / 12	16 / 15-15½	215-225	Creamed	Unmangled
Blue and white checked linen for ordinary covers and bolster cases	84	¾ bleached tow, Fast indigo-dyed tow / ¾ bleached tow, Fast indigo-dyed tow	18, 20 / 20, 20	16 / 17-18	235-245	Yarn-white and dyed	Unmangled
Blue and white striped drill for invalids' coats and trousers	72	¾ bleached flax, Fast indigo-dyed flax / ¾ bleached flax	20, 20 / 18	24 / 18	245-255	Yarn-white and dyed	Unmangled
White linen for handkerchiefs for hospital patients	50/51	4/4 bleached flax / 4/4 bleached flax	30	20-21 / 19-20	102-105	Yarn-white	Lightly mangled
White linen with interwoven red stripes for hospital patients suffering from infectious diseases	50/51	4/4 bleached flax / 4/4 bleached flax	30	20-21 / 19-20	100-105	Yarn-white	Lightly mangled

B. COTTONS.

Stuff.	Width in cm.	Kind of Yarn.	Yarn No.	No. of threads per 1 sq. cm.	Weight per running metre in grams.	Finish. Whether unbleached, creamed, re-bleached or dyed.	Whether and to what extent mangled.
Fustian for lining invalids' coats	81	Cotton	16-71 / 3-4	13-15 / 17-19	310-320	Unbleached	Raised
Bleached fustian for vests	77	Cotton	16-17 / 3-4	14-15 / 17-19	260-270	Bleached	Raised
Calico for drawers	75/76	Cotton	18-19 / 14-15	21-23 / 17-19	190-200	Unbleached	Unmangled
Calico for invalids' shirts	76	Cotton	16-17 / 16-17	24-26 / 25-27	130-140	Bleached	Lightly mangled
Cotton cloth for neckcloths	83	Cotton	33-34 / 42-44	31-33 / 29-31	110-120	Bleached	Dressed
Blue and white checked cottons for ordinary bed and pillow covers	84	¾ bleached Fast indigo-dyed / ¼ bleached Fast indigo-dyed	16 / 16	23-24 / 23-24	152-162	Yarn-white and dyed	Unmangled

C. WOOLLENS (FLANNEL).

Stuff.	Width in cm.	Kind of Yarn.	Yarn No.	No. of threads per 1 sq. cm.	Weight per running metre in grams.	Finish.	Whether and to what extent mangled.
Flannel for bandages	125	Carded yarn of German wool / do.	11-11½ / 8-8½	20-21 / 15-16	320-330	Washed, fulled, lightly blued	Evenly raised on both sides

The remarks on p. 190 apply equally in this case. For stuffs to be worked up in a shrunk state the loss in shrinking must be borne in mind.

The regulations for taking delivery given on p. 191 also apply. The fastness of the blue checks must be tested by four or five washings.

CONTRACT SPECIFICATIONS FOR FRENCH ARMY CLOTHING.

The new regulations issued in 1893 deal with the raw material, dyeing, fulling and dressing of the cloth.

The raw material must consist of wool out of the centre of the fleece, partly of French origin and partly from Morocco, Algiers and Argentina, corresponding in staple to the type adopted by the Ministry of War. The wool must be carefully sorted, cleaned and washed. The use of lamb's wool, skin wool, combings, shoddy, waste wool or cotton is strictly prohibited. Cleaning may be effected by the old or new (carbonisation) process, but in the latter case the strength of the acid used must not exceed 5° B.

Scarlet, jonquil-yellow and maroon are piece-dyed, for which end the goods may be previously bleached with sulphurous acid, bisulphite, or hydrogen peroxide, but all other colours must be dyed in the wool. Cochineal or lac dye is prescribed for scarlet; wood for yellow; sandal wood and madder for maroon. All shades of blue and grey must be vat-dyed, and must not contain any naturally yellow or dark wool. Red may be dyed with madder or artificial alizarine, but the War Ministry reserves the right to prescribe more particularly in each individual case. For black the vat-dyed grey is to be finished with mineral salts and substances containing tannin. Only the very dark blue for non-commissioned officers' uniforms may have a slight addition of sandal wood (but no campeachy) with ferrous sulphate and sumac.

The use of modern dye stuffs is not prohibited so far as they are found to correspond in fastness to the requirements laid down.

The wool after willowing must be carded three or four times and the yarn must be nicely and evenly twisted.

Power-loom weaving alone is admissible, hand weaving being disallowed. The woven fabric may be freed from fat either before or after fulling, according to the softening employed. After fulling, the cloth must not be artificially widened by stretching on the drying frame, and the sprinkling of the goods with glycerine, dissolved glue or vegetable mucilage is prohibited. Yellow and scarlet-coloured cloths must be decatired by steam, but they may be delivered without if the manufacturer is not installed for carrying out that operation.

Length.—The pieces after steaming must not be longer than 40 metres (43¾ yds.) nor less than 25 metres (27⅜ yds.), and the *width* between 137 and 143 *cm.* (54 to 56½ in.).

The *selvedges*, composed of twelve warp threads, must be 18 to 24 *mm.* (¾ to 1 in. nearly) wide after fulling. In the wool-dyed cloths they are to consist of white threads, otherwise of three black and nine threads of the ground colour of the cloth.

In testing the weight of the cloth an average of 12·5 per cent. of moisture is considered normal (at 15° C. and in an atmosphere containing 80 to 90 per cent. of moisture). The weight per metre of cloth for non-commissioned officers' wear, 140 *cm.* (55 in.) wide, exclusive of the selvedges, is fixed at 720 grams (23¼ oz. per yd.) and the minimum breaking strain at 30 kilos. (66 lbs.) for the warp and 22 kilos. (48·4 lbs.) for the weft.

INDEX.

A.

Alkalis, 47, 49.
Alpaca, artificial, 33.
— wool, 31.
Ammoniacal copper oxide, 44.
Analysis of tissues, 147.
Aniline sulphate solution, 47.
Arsenic, estimating, 178.
Artificial wools, 33.
Atlas twills, 119, 120.

B.

Bast fibres, 14.
Batiste, 120.
Black dyes, 178.
Blue dyes, 169.
Breaking length, 97.
— strain of cloth, 94.
— — testers, 128.
Brocade, 121.
Brown dyes, 129.
Buckskin, 121.

C.

Calico, 120.
Camel hair, 32
Cashmere wool, 31.
Cassinet, 122.
Cellulose, 15.
Chappe silk, 41.
— — reeling, 71.
Chiffon, 120.
Cloth testing, 118.
— weighing, 144.
Cocoanut fibre, 26.
Colorations, microchemical, 48.
Colours, fastness of, 158.
Combination of threads in weaving, 124.

Conditioning apparatus, 110-117.
Cordonnet, 89.
Cosmos fibre, 26.
Cotton, 16.
— dead, 17.
— yarn, numbering, 58.
— — twist of, 84.
Count of cloth, 141.
Cows and calves' hair, 33.
Creas, 120.

D.

Dissecting microscope, 9.
Double twills, 119.
Dye, estimation of, 168.
Dynamometers, 128.

E.

Elasticity of cloth, 125.
— silk, 96.
— yarn, 93, 96.
— tester, 99, 130, 136, 138.
Embroidery silk, 89.
Examination of mixed fibres, table of, 51.
Extract wool, 33.

F.

Fastness against rain, 161.
— — street mud, 161.
— to air, 162.
— to weather, 162.
— under friction, 160.
— — washing, 159.
(*See* also Resistance.)
Fat, percentage in yarn, 107.
Figured fabrics, 119.
Findeisen's conditioning apparatus, 117.

Flax, 18.
Floconné, 121.
Fuchsine solution, 46.

G.

German army clothing specifications, 180.
Glover's wool, 30.
Green dyes, 170.
Grège, 88.
Grey dyes, 172.

H.

Hare and rabbit fur, 32.
Heal's conditioning apparatus, 114.
Hemp, 20.
— canvas, 122.
Horse hair, 33.
Hygroscopicity, determination of, 157.
Hyparchia janira, 8.

I.

Imitation yarn, 1.
Indigo dyes, 169.
Iodine solution, 46.
Ironing, resistance to, 163.

J.

Jute, 23.
— fabrics, 120.
— yarns, numbering, 64.

K.

Knapsack cloths, 196.
Kohl's conditioning apparatus, 113.

L.

Linens, 121.
— and cottons for military use, 187.
Linen yarns, numbering, 62.
— — twist of, 86.
Llama wool, 31.

M.

Manila hemp, 25.
Marabout silk, 89.
Measuring machines, 164.
Micrometer, 10.
Microscope, 3-12.
— eye-piece, 4, 7.
— objective, 4, 7.
— judging, 8.
— test objects, 8.
Mode colours, 172.
Modes of weaving, testing, 124.
Mohair, 122.
— wool, 30.
Moisture, determination of, 108.
Mollinos, 120.
Mordants, estimation of, 165.
Mungo, 33.

N.

Naphthol solution, 47.
Nettle fibre, 22.
— yarn, numbering, 65.
New Zealand flax, 26.
Nickel solution, 46.

O.

Organzine, 89.
Orleans, 122.

P.

Packing cloths, 120.
Percale, 120.
Phloroglucin solution, 47.
Piece goods, length of, 164.
Pleurosigma angulatum, 9.
Poil silk, 89.
Polarising apparatus, 11.
Poplin, 122.
Preparing sections, 13.
Provision bags, cloth for, 196.

Q.

Quality numbers, 93.
Quantitative analysis of fabrics, 110.

INDEX.

R.

Ramie, 22.
Raw silk, 88.
Reeling, various standards of, 57-72.
Repp, 122.
Reprise, 109.
Resistance to ironing, 163.
— — perspiration, 160.
— — steaming, 163.

S.

Salts, solutions of, 48.
Sampling reel, 77.
Semi-taffeta, 122.
Separation of wool from cotton, etc., 52-54.
Serge, 121.
Sewing silk, 89.
Sheep's wool, 29.
Shoddy yarn, 1.
Shoe canvas, 192.
Shotting, count of, 141.
Shrinkage, 143.
Silk, 39.
— artificial, 43.
— cordonnet, 89.
— loaded, 42.
— loading, 55.
— spun, 41.
— wild, 41.
Sodium copper oxide, 45.
Steaming, resistance to, 163.
Stiffened linens, 120.
Sulphuric acid, 46, 47, 50.

T.

Tanner's wool, 30.
Tenacity, 127.
Tent canvas, 199.
Thibet wool, 31.
Thickness of cloth, 143.
Thread counter, 125.
Tussah silk, 41.

Twills, 119, 120.
Twist, 1.
— counter, 91.
— tester, 90.

U.

Ulmann's conditioning apparatus, 115, 116.
Unions, 122.

V.

Vegetable fibres, 14.
Velvety fabrics, 120.
Vicuña wool, 31.
— yarn, 1.

W.

Waterproof cloth for troops, 192.
— testing, 154.
Weavings, plain or smooth, 89.
Wool, kinds of, 27.
— muslin, 121.
Woollen yarn, numbering, 66.
— — twist of, 86.

Y.

Yarn, balances for weighing, 72-77.
— doubled, 2.
— external appearance of, 81.
— length of, 79.
— number, determining, 57.
— tensile strength of, 93.
— testers, 81, 97-106.
— testing, 2.
— twisted, 2.
— uniformity of, 94.
Yellow and orange dyes, 172.

Z.

Zanella, 121.
Zinc chloride solution, 46.

ABERDEEN UNIVERSITY PRESS.

A CATALOGUE

OF

Special Technical Works

FOR

Manufacturers, Professional Men, Students, Colleges and Technical Schools

BY EXPERT WRITERS

FOR THE

Oil, Grease, Paint, Colour, Varnish, Soap, Candle, Chemical, Textile, Leather, Pottery, Glass, Plumbing and Decorating Trades and Scientific Professions.

PUBLISHED BY

SCOTT, GREENWOOD & CO.,

TECHNICAL LITERATURE AND TRADE JOURNAL EXPERTS,

19 LUDGATE HILL, LONDON, E.C.

Telegraphic Address: "PRINTERIES, LONDON". Telephone No. 5403, Bank.

N.B.—*Full Particulars of Contents of any of the following books sent post free on application.*

Books on Oils, Soaps, Colours, Glue, Varnishes, etc.

THE PRACTICAL COMPOUNDING OF OILS, TALLOW AND GREASE FOR LUBRICATION, ETC. By AN EXPERT OIL REFINER. Price: United Kingdom, 7s. 6d.; Continent, 9s., post free.

Contents.

Chapters I., **Introductory Remarks** on the General Nomenclature of Oils, Tallow and Greases suitable for Lubrication.—II., **Hydrocarbon Oils.**—III., **Animal and Fish Oils.**—IV., **Compound Oils.**—V., **Vegetable Oils.**—VI., **Lamp Oils.**—VII., **Engine Tallow, Solidified Oils and Petroleum Jelly.**—VIII., **Machinery Greases: Loco and Antifriction.**—IX., **Clarifying and Utilisation of Waste Fats, Oils, Tank Bottoms, Drainings of Barrels and Drums, Pickings Up, Dregs, etc.**—X., **The Fixing and Cleaning of Oil Tanks, etc.**—Appendix of General Information.

Press Opinions.

"This work is written from the standpoint of the oil trade, but its perusal will be found very useful by users of machinery and all who have to do with lubricants in any way."—*Colliery Guardian.*

"The properties of the different grades of mineral oil and of the animal and vegetable non-drying oils are carefully described, and the author justly insists that the peculiarities of the machinery on which the lubricants are to be employed must be considered almost before everything else. . . . The chapters on grease and solidified oils, etc., are excellent."—*The Ironmonger.*

"In its ninety-six pages this little work contains a wealth of information; it is written without waste of words on theoretical matters, and contains numerous formulas for a great variety of compounds for the most varied lubricants. In addition there are many practical hints of use in the factory in general, such as of tanks, etc., and altogether the book is worth several times its price in any factory of these compounds."—*American Soap Journal.*

SOAPS. A Practical Manual of the Manufacture of Domestic, Toilet and other Soaps. By GEORGE H. HURST, F.C.S. Illustrated with 66 Engravings. Price 12s. 6d.; Germany, 14 mks.; France and Belgium, 16 frs., post free.

Contents.

Chapters I., **Introductory.**—II., **Soap-maker's Alkalies.**—III., **Soap Fats and Oils.**—IV., **Perfumes.**—V., **Water as a Soap Material.**—VI., **Soap Machinery.**—VII., **Technology of Soap-making.**—VIII., **Glycerine in Soap Lyes.**—IX., **Laying out a Soap Factory.**—X., **Soap Analysis.**—Appendices.

Press Opinions.

"Much useful information is conveyed in a convenient and trustworthy manner which will appeal to practical soap-makers."—*Chemical Trade Journal.*

"This is a better book on soap-manufacture than any of the same size which have been published for some time. It reads like the 'real thing,' and gives a very complete account of the technique of soap-making, especially of the machinery employed, the different methods and even the arrangement of soap factories. . . . The book is produced well, and is splendidly illustrated "—*Chemist and Druggist.*

"The best and most reliable methods of analysis are fully discussed, and form a valuable source of reference to any work's chemist. . . . Our verdict is a capitally-produced book, and one that is badly needed."—*Birmingham Post.*

"We think it is the most practical book on these subjects that has come to us from England so far."—*American Soap Journal.*

"Works that deal with manufacturing processes, and applied chemistry in particular, are always welcome. Especially is this the case when the material presented is so up-to-date as we find it here."—*Bradford Observer.*

ANIMAL FATS AND OILS: Their Practical Production, Purification and Uses for a Great Variety of Purposes. Their Properties, Falsification and Examination. A Handbook for Manufacturers of Oil and Fat Products, Soap and Candle Makers, Agriculturists, Tanners, Margarine Manufacturers, etc., etc. By LOUIS EDGAR ANDÉS. With 62 Illustrations. Price 10s. 6d.; France and Belgium, 13 frs.; Colonies, 12s., post free.

Contents.

Introduction. Occurrence, Origin, Properties and Chemical Constitution of Animal Fats. Preparation of Animal Fats and Oils. Machinery. Tallow-melting Plant. Extraction Plant. Presses. Filtering Apparatus. Butter: Raw Material and Preparation, Properties, Adulterations, Beef Lard or Remelted Butter, Testing. Candle-fish Oil. Mutton Tallow. Hare Fat. Goose Fat. Neatsfoot Oil. Bone Fat: Bone Boiling, Steaming Bones, Extraction, Refining. Bone Oil. Artificial Butter: Oleomargarine, Margarine Manufacture in France, Grasso's Process. "Kaiser's Butter," Jahr & Münzberg's Method, Filbert's Process, Winter's Method. Human Fat. Horse Fat. Beef Marrow. Turtle Oil. Hog's Lard: Raw Material, Preparation, Properties, Adulterations, Examination. Lard Oil. Fish Oils. Liver Oils. Artificial Train Oil. Wool Fat: Properties, Purified Wool Fat. Spermaceti: Examination of Fats and Oils in General.

Press Opinions.

"The latest and most improved forms of machinery are in all cases indicated, and the many advances which have been made during the past years in the methods of producing the more common animal fats—lard, tallow and butter—receive due attention."—*Glasgow Herald.*

"The work is very fully illustrated, and the style throughout is in strong contrast to that employed in many such treatises, being simple and clear."— *Shoe and Leather Record*

"An important handbook for the 'fat industry,' now a large one. The explanation of the most scientific processes of production lose nothing of their clearness in the translation."—*Newcastle Chronicle.*

"It is a valuable work, not only for the student, but also for the practical manufacturer of oil and fat products.."—*Journal of the American Chemical Society.*

"The descriptions of technical processes are clear, and the book is well illustrated and should prove useful."—*Manchester Guardian.*

VEGETABLE FATS AND OILS: Their Practical Preparation,
Purification and Employment for Various Purposes, their Properties, Adulteration and Examination. A Handbook for Oil Manufacturers and Refiners, Candle, Soap and Lubricating Oil Makers, and the Oil and Fat Industry in General. Translated from the German of LOUIS EDGAR ANDÉS. With 94 Illustrations. Price 10s. 6d.; Germany, 12 mks.; France and Belgium, 13 frs.; Colonies, 12s. post free.

Contents.

Statistical Data. General Properties of the Vegetable Fats and Oils. Estimation of the Amount of Oil in Seeds. Table of Vegetable Fats and Oils, with French and German Nomenclature, Source and Origin and Percentage of Fat in the Plants from which they are Derived. The Preparation of Vegetable Fats and Oils: Storing Oil Seeds; Cleaning the Seed. Apparatus for Grinding Oil Seeds and Fruits. Installation of Oil and Fat Works. Extraction Method of Obtaining Oils and Fats. Oil Extraction Installations. Press Moulds. Non-drying Vegetable Oils. Vegetable Drying Oils. Solid Vegetable Fats. Fruits Yielding Oils and Fats. Wool-softening Oils. Soluble Oils. Treatment of the Oil after Leaving the Press. Improved Methods of Refining with Sulphuric Acid and Zinc Oxide or Lead Oxide. Refining with Caustic Alkalies, Ammonia, Carbonates of the Alkalies, Lime. Bleaching Fats and Oils. Practical Experiments on the Treatment of Oils with regard to Refining and Bleaching. Testing Oils and Fats.

Press Opinions.

"Concerning that and all else within the wide and comprehensive connexion involved, this book must be invaluable to every one directly or indirectly interested in the matters it treats of."—*Commerce.*

"The proprietors of the *Oil and Colourman's Journal* have not only placed a valuable and highly interesting book of reference in the hands of the fats and oils industry in general, but have rendered no slight service to experimental and manufacturing chemists."—*Manufacturing Chemist.*

LUBRICATING OILS, FATS AND GREASES: Their Origin,
Preparation, Properties, Uses and Analyses. A Handbook for Oil Manufacturers, Refiners and Merchants, and the Oil and Fat Industry in General. By GEORGE H. HURST, F.C.S. Price 10s. 6d.; Germany, 12 mks.; France and Belgium, 13 frs.; Colonies, 12s., post free.

Contents.

Chapters I., **Introductory.** Oils and Fats, Fatty Oils and Fats, Hydrocarbon Oils, Uses of Oils.—II., **Hydrocarbon Oils.** Distillation, Simple Distillation, Destructive Distillation, Products of Distillation, Hydrocarbons, Paraffins, Olefins, Naphthenes.—III., **Scotch Shale Oils.** Scotch Shales, Distillation of Scotch Oils, Shale Retorts, Products of Distilling Shales, Separating Products, Treating Crude Shale Oil, Refining Shale Oil, Shale Oil Stills, Shale Naptha Burning Oils, Lubricating Oils, Wax.—IV., **Petroleum.** Occurrence, Geology, Origin, Composition, Extraction, Refining, Petroleum Stills, Petroleum Products, Cylinder Oils, Russian Petroleum, Deblooming Mineral Oils.—V., **Vegetable and Animal Oils.** Introduction, Chemical Composition of Oils and Fats, Fatty Acids, Glycerine, Extraction of Animal and Vegetable Fats and Oils, Animal Oils, Vegetable Oils, Rendering, Pressing, Refining, Bleaching, Tallow, Tallow Oil, Lard Oil, Neatsfoot Oil, Palm Oil, Palm Nut Oil, Cocoanut Oil, Castor Oil, Olive Oil, Rape and Colza Oils, Arachis Oil, Niger Seed Oil, Sperm Oils, Whale Oil, Seal Oil, Brown Oils, Lardine, Thickened Rape Oil.—VI., **Testing and Adulteration of Oils.** Specific Gravity, Alkali Tests, Sulphuric Acid Tests, Free Acids in Oils, Viscosity Tests, Flash and Fire Tests, Evaporation Tests, Iodine and Bromide Tests, Elaidin Test, Melting Point of Fat, Testing Machines.—VII., **Lubricating Greases.** Rosin Oil, Anthracene Oil, Making Greases, Testing and Analysis of Greases.—VIII., **Lubrication.** Friction and Lubrication, Lubricant, Lubrication of Ordinary Machinery, Spontaneous Combustion of Oils, Stainless Oils, Lubrication of Engine Cylinders, Cylinder Oils.—**Appendices.** A. Table of Baume's Hydrometer—B. Table of Thermometric Degrees—C. Table of Specific Gravities of Oils.—**Index.**

Press Opinions.

"This is a clear and concise treatment of the method of manufacturing and refining lubricating oils. . . . The book is one which is well worthy the attention of readers who are users of oil."—*Textile Recorder.*

"The book is well printed, and is a credit alike to author, printer and publisher."—*Textile Mercury.*

"Mr. Hurst has in this work supplied a practical treatise which should prove of especial value to oil dealers and also, though in a less degree, of oil users."—*Textile Manufacturer.*

"A mere glance at the table of contents is sufficient to show how various are the conditions to which these materials have to be applied, how much knowledge is required for the selection of the right kind for each particular purpose, and how by processes of mixture or manufacture the requisite qualities are obtained in each case."—*Manchester Guardian.*

"This valuable and useful work, which is both scientific and practical, has been written with a view of supplying those who deal in and use oils, etc., for the purpose of lubrication with some information respecting the special properties of the various products which cause these various oils to be of value as lubricants."—*Industries and Iron.*

"We have no hesitation in saying that in our opinion this book ought to be very useful to all those who are interested in oils, whether as manufacturers or users of lubricants, or to those chemists or engineers whose duty it may be to report upon the suitability of the same for any particular class of work."—*Engineer.*

"The author is widely known and highly respected as an authority on the chemistry of oils and the technics of lubrication, and it is safe to say that no work of similar interest or equal value to the general oil-selling and consuming public has heretofore appeared in the English language."—*Drugs, Oils and Paints,* U.S.A.

"It will be a valuable addition to the technical library of every steam user's establishment."—*Machinery Market.*

THE MANUFACTURE OF VARNISHES, OIL REFINING AND BOILING, AND KINDRED INDUSTRIES.

Describing the Manufacture of Spirit Varnishes and Oil Varnishes; Raw Materials: Resins, Solvents and Colouring Principles; Drying Oils: their Properties, Applications and Preparation by both Hot and Cold Processes; Manufacture, Employment and Testing of Different Varnishes. Translated from the French of ACH. LIVACHE. Greatly Extended and Adapted to English Practice, with numerous Original Recipes. By J. G. McINTOSH, Lecturer on Oils, Colours and Varnishes. Price 12s. 6d. France and Belgium, 16 frs.; Colonies, 14s., post free.

Contents.

I. Resins: Gum Resins, Oleo Resins and Balsams, Commercial Varieties, Source, Collection, Characteristics, Chemical Properties, Physical Properties, Hardness, Adulterations, Appropriate Solvents, Special Treatment, Special Use.—II. Solvents: Natural, Artificial, Manufacture, Storage, Special Use.—III. Colouring: Principles, (1) Vegetable, (2) Coal Tar, (3) Coloured Resinates, (4) Coloured Oleates and Linoleates.—Gum Running: Furnaces, Bridges, Flues, Chimney Shafts, Melting Pots, Condensers, Boiling or Mixing Pans, Copper Vessels, Iron Vessels (Cast), Iron Vessels (Wrought), Iron Vessels (Silvered), Iron Vessels (Enamelled), Steam Superheated Plant, Hot-air Plant.—Spirit Varnish Manufacture: Cold Solution Plant, Mechanical Agitators, Hot Solution Plant, Jacketted Pans, Mechanical Agitators, Clarification and Filtration, Bleaching Plant, Storage Plant.—Manufacture, Characteristics and Uses of the Spirit Varnishes yielded by: Amber, Copal, Dammar, Shellac, Mastic, Sandarac, Rosin, Asphalt, India Rubber, Gutta Percha, Collodion, Celluloid, Resinates, Oleates.—Manufacture of Varnish Stains.—Manufacture of Lacquers.—Manufacture of Spirit Enamels.—Analysis of Spirit Varnishes.—Physical and Chemical Constants of Resins.—Table of Solubility of Resins in different Menstrua.—Systematic qualitative Analysis of Resins Hirschop's tables.—Drying Oils: Oil Crushing Plant, Oil Extraction Plant, Individual Oils, Special Treatment of Linseed Oil, Poppyseed Oil, Walnut Oil, Hempseed Oil, Llamantia Oil, Japanese Wood Oil, Gurjun Balsam, Climatic Influence on Seed and Oil.—Oil Refining: Processes, Thenard's, Liebig's, Filtration, Storage, Old Tanked Oil.—Oil Boiling: Fire Boiling Plant, Steam Boiling Plant, Hot-air Plant, Air Pumps, Mechanical Agitators, Vincent's Process, Hadfield's Patent, Storer's Patent, Walton's Processes, Continental Processes, Pale Boiled Oil, Double Boiled Oil, Hartley and Blenkinsop's Process.—Driers: Manufacture, Special Individual Use of (1) Litharge, (2) Sugar of Lead, (3) Red Lead, (4) Lead Borate, (5) Lead Linoleate, (6) Lead Resinate, (7) Black Oxide of Manganese, (8) Manganese Acetate, (9) Manganese Borate, (10) Manganese Resinate, (11) Manganese Linoleate, Mixed Resinates and Linoleates, Manganese and Lead, Zinc Sulphate, Terebine, Liquid Driers.—Solidified Boiled Oil.—Manufacture of Linoleum.—Manufacture of India Rubber Substitutes.—Printing Ink Manufacture.—Lithographic Ink Manufacture.—Manufacture of Oil Varnishes.—Running and Special Treatment of Amber, Copal, Kauri, Manilla.—Addition of Oil to Resin.—Addition of Resin to Oil.—Mixed Processes.—Solution in Cold of previously fused Resin.—Dissolving Resins in Oil, etc., under pressure.—Filtration.—Clarification.—Storage.—Ageing.—Coachmakers' Varnishes and Japans.—Oak Varnishes.—Japanners' Stoving Varnishes.—Japanners' Gold Size.—Brunswick Black.—Various Oil Varnishes.—Oil-Varnish Stains.—Varnishes for "Enamels".—India Rubber Varnishes.—Varnishes Analysis: Processes, Matching.—Faults in Varnishes: Cause, Prevention.—Experiments and Exercises.

Press Opinions.

"There is no question that this is a useful book."—*Chemist and Druggist.*

"The different formulæ which are quoted appear to be far more 'practical' than such as are usually to be found in text-books; and assuming that the original was published two or three years ago, and was only slightly behindhand in its information, the present volume gives a fair insight into the position of the varnish industry."—*The Ironmonger.*

THE TESTING AND VALUATION OF RAW MATERIALS USED IN PAINT AND COLOUR MANUFACTURE.

By M. W. JONES, F.C.S. A book for the laboratories of colour works. Price 5s.; Colonies and Continent, 6s., strictly net, post free.

Contents.

Aluminium Compounds. China Clay. Iron Compounds. Potassium Compounds. Sodium Compounds. Ammonium Hydrate. Acids. Chromium Compounds. Tin Compounds. Copper Compounds. Lead Compounds. Zinc Compounds. Manganese Compounds. Arsenic Compounds. Antimony Compounds. Calcium Compounds. Barium Compounds. Cadmium Compounds. Mercury Compounds. Ultramarine. Cobalt and Carbon Compounds. Oils. Index.

THE CHEMISTRY OF ESSENTIAL OILS AND ARTIFICIAL PERFUMES. By ERNEST J. PARRY, B.Sc. (Lond.), F.I.C., F.C.S. Illustrated with 20 Engravings. 400 pp. Price 12s. 6d.; Abroad, 14s., strictly net, post free.

Contents.

Chapters I., **The General Properties of Essential Oils.**—II., **Compounds occurring in Essential Oils.**—III., **The Preparation of Essential Oils.**—IV., **The Analysis of Essential Oils.**—V., **Systematic Study of the Essential Oils.**—VI., **Terpeneless Oils.**—VII., **The Chemistry of Artificial Perfumes.**—Appendix: Table of Constants.

Press Opinions.

"At various times monographs have been printed by individual workers, but it may safely be said that Mr. Parry is the first in these latter days to deal with the subject in an adequate manner. His book is well conceived and well written. . . . He is known to have sound practical experience in analytical methods, and he has apparently taken pains to make himself *au fait* with the commercial aspects of the subject."—*Chemist and Druggist.*

"We can heartily recommend this volume to all interested in the subject of essential oils from the scientific or the commercial standpoint."—*British and Colonial Druggist.*

"There can be no doubt that the publication will take a high place in the list of scientific text-books."—*London Argus.*

"A most useful appendix is inserted, giving a table of constants for the more important essential oils. . . . This, in itself, is of sufficient importance and use to warrant the publication of the book, and, added to the very complete classification and consideration of the essential oils which precedes it, the volume becomes of great value to all interested."—*Glasgow Herald.*

"Mr. Parry has done good service in carefully collecting and marshalling the results of the numerous researches published in various parts of the world."—*Pharmaceutical Journal.*

COLOUR: A HANDBOOK OF THE THEORY OF COLOUR. By GEORGE H. HURST, F.C.S. With 10 coloured Plates and 72 Illustrations. Price 7s. 6d.; Abroad, 9s., post free.

Contents.

Chapters I., **Colour and its Production.**—II., **Cause of Colour in Coloured Bodies.**—III., **Colour Phenomena and Theories.**—IV., **The Physiology of Light.**—V., **Contrast.**—VI., **Colour in Decoration and Design.**—VII., **Measurement of Colour.**

Press Opinions.

"This is a workmanlike technical manual, which explains the scientific theory of colour in terms intelligible to everybody. . . . It cannot but prove both interesting and instructive to all classes of workers in colour."—*Scotsman.*

"Mr. Hurst's *Handbook on the Theory of Colour* will be found extremely useful, not only to the art student, but also to the craftsman, whose business it is to manipulate pigments and dyes."—*Nottingham Daily Guardian.*

"It is thoroughly practical, and gives in simple language the why and wherefore of the many colour phenomena which perplex the dyer and the colourist."—*Dyer and Calico Printer.*

"We have found the book very interesting, and can recommend it to all who wish to master the different aspects of colour theory, with a view to a practical application of the knowledge so gained."—*Chemist and Druggist.*

"It will be found to be of direct service to the majority of dyers, calico printers and colour mixers, to whom we confidently recommend it."—*Chemical Trade Journal.*

"This useful little book possesses considerable merit, and will be of great utility to those for whom it is primarily intended."—*Birmingham Post.*

IRON-CORROSION, ANTI-FOULING AND ANTI-CORROSIVE PAINTS. By LOUIS EDGAR ANDÉS. Sixty-two Illustrations. Translated from the German. Price 10s. 6d.; Abroad, 12s., strictly net, post free.

Contents.

Ironrust and its Formation—Protection from Rusting by Paint—Grounding the Iron with Linseed Oil, etc.—Testing Paints—Use of Tar for Painting on Iron—Anti-corrosive Paints—Linseed Varnish—Chinese Wood Oil—Lead Pigments—Iron Pigments—Artificial Iron Oxides—Carbon—Preparation of Anti-corrosive Paints—Results of Examination of Several Anti-corrosive Paints—Paints for Ship's Bottoms—Anti-fouling Compositions—Various Anti-corrosive and Ship's Paints—Official Standard Specifications for Ironwork Paints—Index.

THE LEATHER WORKER'S MANUAL. Being a Compendium of Practical Recipes and Working Formulæ for Curriers, Bootmakers, Leather Dressers, Blacking Manufacturers, Saddlers, Fancy Leather Workers, and all Persons engaged in the Manipulation of Leather. By H. C. STANDAGE. Price 7s. 6d.; Abroad, 9s., strictly net, post free.

Contents.

Chapters I., Blackings, Polishes, Glosses, Dressings, Renovators, etc., for Boot and Shoe Leather.—II., Harness Blackings, Dressings, Greases, Compositions, Soaps, and Boot-top Powders and Liquids, etc., etc.—III., Leather Grinders' Sundries.—IV., Currier's Seasonings, Blacking Compounds, Dressings, Finishes, Glosses, etc.—V., Dyes and Stains for Leather.—VI., Miscellaneous Information.—VII., Chrome Tannage.—Index.

Press Opinions.

"The book being absolutely unique, is likely to be of exceptional value to all whom it concerns, as it meets a long-felt want."—*Birmingham Gazette.*

"This is a valuable collection of practical receipts and working formulæ for the use of those engaged in the manipulation of leather. We have no hesitation in recommending it as one of the best books of its kind, an opinion which will be endorsed by those to whom it appeals."—*Liverpool Mercury.*

GLUE AND GLUE TESTING. By SAMUEL RIDEAL, D.Sc. Lond., Fellow of the Institute of Chemistry, Vice-President of the Society of Public Analysts, Author of "Water and its Purification," "Disinfection and Disinfectants". Illustrated with fourteen Engravings. Price 10s. 6d., strictly net; United States, 3 dols.; Germany, 12 mks.; France and Belgium, 13 frs., post free.

Contents.

Chapters I., **Constitution and Properties**: Definitions, Sources, Gelatine, Chondrin and Allied Bodies, Physical and Chemical Properties, Classification, Grades and Commercial Varieties.—II., **Raw Materials and Manufacture**: Glue Stock, Lining, Extraction, Washing and Clarifying, Filter Presses, Water Supply, Use of Alkalies, Action of Bacteria and of Antiseptics, Various Processes, Cleansing, Forming, Drying, Crushing, etc., Secondary Products.—III., **Uses of Glue**: Selection and Preparation for Use, Carpentry, Veneering, Paper Making, Book-binding, Printing Rollers, Hectographs, Match Manufacture, Sandpaper, etc., Substitutes for other Materials, Artificial Leather and Caoutchouc.—IV., **Gelatine**: General Characters, Liquid Gelatine, Photographic Uses, Size, Tanno- Chrome, and Formo-Gelatine, Artificial Silk, Cements, Pneumatic Tyres, Culinary, Meat Extracts, Isinglass, Medicinal and other Uses, Bacteriology.—V., **Glue Testing**: Review of Processes, Chemical Examination, Adulteration, Physical Tests, Valuation of Raw Materials.—VII., **Commercial Aspects.**

Books on Pottery, Glass, etc.

THE MANUAL OF PRACTICAL POTTING. Price 17s. 6d.; Colonies and Continent, 18s., post fre:.

Contents.

Introduction. The Rise and Progress of the Potter's Art.—Chapters I., **Bodies.** China and Porcelain Bodies, Parian Bodies, Semi-porcelain and Vitreous Bodies, Mortar Bodies, Earthenwares Granite and C.C. Bodies, Miscellaneous Bodies, Sagger and Crucible Clays, Coloured Bodies, Jasper Bodies, Coloured Bodies for Mosaic Painting, Encaustic Tile Bodies, Body Stains, Coloured Dips.—II., **Glazes.** China Glazes, Ironstone Glazes, Earthenware Glazes, Glazes without Lead, Miscellaneous Glazes, Coloured Glazes, Majolica Colours.—III., **Gold and Cold Colours.** Gold, Purple of Cassius, Marone and Ruby, Enamel Colour Bases, Enamel Colour Fluxes, Enamel Colours, Mixed Enamel Colours, Antique and Vellum Enamel Colours, Underglaze Colours, Underglaze Colour Fluxes, Mixed Underglaze Colours, Flow Powders, Oils and Varnishes.—IV., **Means and Methods.** Reclamation of Waste Gold, The Use of Cobalt, Notes on Enamel Colours, Liquid or Bright Gold.—V., **Classification and Analysis.** Classification of Clay Ware, Lord Playfair's Analysis of Clays, The Markets of the World, Time and Scale of Firing, Weights of Potter's Material, Decorated Goods Count.—VI., Comparative Loss of Weight of Clays.—VII., Ground Felspar Calculations.—VIII., The Conversion of Slop Body Recipes into Dry Weight.—IX., The Cost of Prepared Earthenware Clay.—X., **Forms and Tables.** Articles of Apprenticeship, Manufacturer's Guide to Stocktaking, Table of Relative Values of Potter's Materials, Hourly Wages Table, Workman's Settling Table, Comparative Guide for Earthenware and China Manufacturers in the Use of Slop Flint and Slop Stone, Foreign Terms applied to Earthenware and China Goods, Table for the Conversion of Metrical Weights and Measures on the Continent of South America.

CERAMIC TECHNOLOGY: Being some Aspects of Technical Science as Applied to Pottery Manufacture. Edited by CHARLES F. BINNS. Price 12s. 6d.; Colonies and Continent, 14s., post free.

Contents.
Preface.—Introduction.—Chapters I., The Chemistry of Pottery—II., Analysis and Synthesis.—III., Clays and their Components.—IV., The Biscuit Oven.—V., Pyrometry.—VI., Glazes and their Composition.—VII., Colours and Colour-making'—Index.

COLOURING AND DECORATION OF CERAMIC WARE. By ALEX. BRONGNIART. With Notes and Additions by ALPHONSE SALVETAT. Translated from the French. The writings of Brongniart marked an epoch in ceramic literature, and are now for the first time offered in book form in English. Any potter or workman who is in any way interested in ceramic ware, glazes or enamels will find this work a perfect mine of information. Hundreds of receipts for making and applying colours, glazes and enamels, firing, etc. Bound in Cloth. 200 pages. Price 7s. 6d., strictly net, post free.

HOW TO ANALYSE CLAY. Practical Methods for Practical Men. By HOLDEN M. ASHBY, Professor of Organic Chemistry. Price 2s. 6d., strictly net, post free.

THE ART OF RIVETING GLASS, CHINA AND EARTHENWARE. By J. HOWARTH. Price 1s.; by post. 1s. 2d.

PAINTING ON GLASS AND PORCELAIN AND ENAMEL PAINTING. A Complete Introduction to the Preparation of all the Colours and Fluxes used for Painting on Porcelain, Enamel, Faience and Stoneware, the Coloured Pastes and Coloured Glasses, together with a Minute Description of the Firing of Colours and Enamels. On the Basis of Personal Practical Experience of the Condition of the Art up to Date. By FELIX HERMANN, Technical Chemist. With 18 Illustrations. Second, greatly Enlarged, Edition. Price 10s. 6d.; Germany, 12 mks.; France and Belgium, 13 frs., post free.

Contents.
History of Glass Painting.—Chapters I., The Articles to be Painted: Glass, Porcelain, Enamel, Stoneware, Faience.—II., Pigments: 1, Metallic Pigments: Antimony Oxide, Naples Yellow, Barium Chromate, Lead Chromate, Silver Chloride, Chromic Oxide.—III., Fluxes: Fluxes, Felspar, Quartz, Purifying Quartz, Sedimentation, Quenching, Borax, Boracic Acid, Potassium and Sodium Carbonates, Rocaille Flux.—IV., Preparation of the Colours for Glass Painting.—V., The Colour Pastes.—VI., The Coloured Glasses.—VII., Composition of the Porcelain Colours.—VIII., The Enamel Colours: Enamels for Artistic Work.—IX., Metallic Ornamentation: Porcelain Gilding, Glass Gilding.—X., Firing the Colours: 1, Remarks on Firing: Firing Colours on Glass, Firing Colours on Porcelain; 2, The Muffle.—XI., Accidents occasionally Supervening during the Process of Firing.—XII., Remarks on the Different Methods of Painting on Glass, Porcelain, etc.—Appendix: Cleaning Old Glass Paintings.

Press Opinions.
"A reliable treatise on the preparation of the colours and fluxes, with exhaustive quantitative recipes, and minute descriptions of the firing of colours and enamels, is of no small technical importance, and emanating from so distinguished an authority as Felix Hermann, Brongniart's successor in the direction of the Sèvres manufactory, merits the earnest study of all engaged in the porcelain and kindred industries in England. . . . In every district of England where art porcelain and glass is manufactured this treatise should be widely circulated, and its contents made familiar to all engaged, in whatever capacity, in the trade."—*Leeds Mercury.*

"The whole cannot fail to be both of interest and service to glass workers and to potters generally, especially those employed upon high-class work."—*Staffordshire Sentinel.*

A Reissue of THE HISTORY OF THE STAFFORDSHIRE POTTERIES; AND THE RISE AND PROGRESS OF THE MANUFACTURE OF POTTERY AND PORCELAIN. With References to Genuine Specimens, and Notices of Eminent Potters. By SIMEON SHAW. (Originally Published in 1829.) Price 7s. 6d., strictly net, post free; Abroad, 9s.

Contents.

Introductory Chapter showing the position of the Pottery Trade at the present time (1899).—Chapters I., **Preliminary Remarks.**—II., **The Potteries,** comprising Tunstall, Brownhills, Greenfield and New Field, Golden Hill, Latebrook, Green Lane, Burslem, Longport and Dale Hall, Hot Lane and Cobridge, Hanley and Shelton, Etruria, Stoke, Penkhull, Fenton, Lane Delph, Foley, Lane End.—III., **On the Origin of the Art,** and its Practice among the early Nations.—IV., **Manufacture of Pottery,** prior to 1700.—V., **The Introduction of Red Porcelain** by Messrs. Elers, of Bradwell, 1690.—VI., **Progress of the Manufacture** from 1700 to Mr. Wedgwood's commencement in 1760.—VII., **Introduction of Fluid Glaze.** Extension of the Manufacture of Cream Colour.—Mr. Wedgwood's Queen's Ware.—Jasper, and Appointment of Potter to her Majesty.—Black Printing.—VIII., **Introduction of Porcelain.** Mr. W. Littler's Porcelain.—Mr. Cookworthy's Discovery of Kaolin and Petuntse, and Patent.—Sold to Mr. Champion—resold to the New Hall Com.—Extension of Term.—IX., **Blue Printed Pottery.** Mr. Turner, Mr. Spode (1), Mr. Baddeley, Mr. Spode (2), Messrs. Turner, Mr. Wood, Mr. Wilson, Mr. Minton.—Great Change in Patterns of Blue Printed.—X., **Introduction of Lustre Pottery.** Improvements in Pottery and Porcelain subsequent to 1800.

Press Opinions.

"This work is all the more valuable because it gives one an idea of the condition of affairs existing in the north of Staffordshire before the great increase in work and population due to modern developments."—*Western Morning News.*

"The book will be especially welcomed at a time when interest in the art of pottery manufacture commands a more widespread and general interest than at any previous time."—*Wolverhampton Chronicle.*

"Copies of the original work are now of considerable value, and the facsimile reprint now issued cannot but prove of considerable interest to all interested in the great industry."—*Derby Mercury.*

"There is much curious and useful information in the work, and the publishers have rendered the public a service in reissuing it."—*Burton Mail.*

A Reissue of THE CHEMISTRY OF THE SEVERAL NATURAL AND ARTIFICIAL HETEROGENEOUS COMPOUNDS USED IN MANUFACTURING PORCELAIN, GLASS, AND POTTERY. By SIMEON SHAW. (Originally Published in 1837.) Price 17s. 6d.; Colonies and Continent, 18s., strictly net, post free.

Contents.

PART I., ANALYSIS AND MATERIALS—Chapters I., **Introduction**: Laboratory and Apparatus; **Elements**: Combinative Potencies, Manipulative Processes for Analysis and Reagents, Pulverisation, Blow-pipe Analysis, Humid Analysis, Preparatory Manipulations, General Analytic Processes, Compounds Soluble in Water, Compounds Soluble only in Acids, Compounds (Mixed) Soluble in Water, Compounds (Mixed) Soluble in Acids, Compounds (Mixed) Insoluble, Particular Analytic Processes—II., **Temperature**: Coal, Steam Heat for Printers' Stoves—III., **Acids and Alkalies**: Boracic Acid, Muriatic Acid, Nitric Acid, Sulphuric Acid, Potash, Soda, Lithia, Calculation of Chemical Separations—IV., **The Earths**: Alumine, Clays, Silica, Flint, Lime, Plaster of Paris, Magnesia, Barytes, Felspar, Grauen (or China Stone) China Clay, Chert—V., **Metals**: Reciprocal Combinative Potencies of the Metals, Antimony, Arsenic, Chromium, Green Oxide, Cobalt, Chromic Acid, Humid Separation of Nickel from Cobalt, Arsenite of Cobalt, Copper, Gold, Iron, Lead, Manganese, Platinum, Silver, Tin, Zinc.

PART II., SYNTHESIS AND COMPOUNDS.—Chapters I., Sketch of the Origin and Progress of the Art —II., **Science of Mixing**: Scientific Principles of the Manufacture, Combinative Potencies of the Earths.—III., **Bodies**: Porcelain—Hard, Porcelain—Fritted Bodies, Porcelain—Raw Bodies, Porcelain—Soft, Fritted Bodies, Raw Bodies, Stone Bodies, Ironstone, Dry Bodies, Chemical Utensils, Fritted Jasper, Fritted Pearl, Fritted Drab, Raw Chemical Utensils, Raw Stone, Raw Jasper, Raw Pearl, Raw Mortar, Raw Drab, Raw Brown Raw Fawn, Raw Cane, Raw Red Porous, Raw Egyptian, Earthenware, Queen's Ware, Cream Colour, Blue and Fancy Printed, Dipped and Mocha, Chalky, Rings, Stilts, etc.—IV., **Glazes**: Porcelain—Hard Fritted, Porcelain—Soft Fritted, Porcelain—Soft Raw, Cream Colour Porcelain, Blue Printed Porcelain, Fritted Glazes, Analysis of Fritt, Analysis of Glaze, Coloured Glazes, Dips, Smears, and Washes; **Glasses**: Flint Glass, Coloured Glasses, Artificial Garnet, Artificial Emerald, Artificial Amethyst, Artificial Sapphire, Artificial Opal, Plate Glass, Crown Glass, Broad Glass, Bottle Glass, Phosphoric Glass, British Steel Glass, Glass-Staining and Painting, Engraving on Glass, Dr. Faraday's Experiments.—V., **Colours**: Colour Making, Fluxes or Solvents, Components of the Colours; **Reds, etc., from Gold**, Carmine or Rose Colour, Purple, Reds, etc., from Iron, Blues, Yellows, Greens, Blacks, White, Silver for Burnishing, Gold for Burnishing, Printer's Oil, Lustres.

PART III., TABLES OF THE CHARACTERISTICS OF CHEMICAL SUBSTANCES.—Preliminary Remarks, Oxygen (Tables), Sulphur and its Compounds, Nitrogen ditto, Chlorine ditto, Bromine ditto, Iodine ditto, Fluorine ditto, Phosphorus ditto, Boron ditto, Carbon ditto, Hydrogen ditto, Observations, Ammonium and its Compounds (Tables), Thorium ditto, Zirconium ditto, Aluminium ditto, Yttrium ditto, Glucinum ditto, Magnesium ditto, Calcium ditto, Strontium ditto, Barium ditto, Lithium ditto, Sodium and its Compounds, Potassium ditto, Observations, Selenium and its Compounds (Tables), Arsenic ditto, Chromium ditto, Vanadium ditto, Molybdenum ditto, Tungsten ditto, Antimony ditto, Tellurium ditto, Tantalum ditto, Titanium ditto, Silicium ditto, Osmium ditto, Gold ditto, Iridium ditto, Rhodium ditto, Platinum ditto, Palladium ditto, Mercury ditto, Silver ditto, Copper ditto, Uranium ditto, Bismuth aud its Compounds, Tin ditto, Lead ditto, Cerium ditto, Cobalt ditto, Nickel ditto, Iron ditto, Cadmium ditto, Zinc ditto, Manganese ditto, Observations, Isomorphous Groups, Isomeric ditto, Metameric ditto, Polymeric ditto, Index.

ENAMELS AND ENAMELLING. An Introduction to the Preparation and Application of all Kinds of Enamels for Technical and Artistic Purposes. For Enamel Makers, Workers in Gold and Silver, and Manufacturers of Objects of Art. By PAUL RANDAU. Translated from the German. With 16 Illustrations. Price ; Abroad, strictly net, post free.

Contents.

I., Introduction.—II., Composition and Properties of Glass.—III., Raw Materials for the Manufacture of Enamels.—IV., Substances Added to Produce Opacity.—V., Fluxes. VI., Pigments.—VII., Decolorising Agents.—VIII., Testing the Raw Materials with the Blow-pipe Flame.—IX., Subsidiary Materials.—X., Preparing the Materials for Enamel Making.—XI., Mixing the Materials.—XII., **The Preparation of Technical Enamels**: The Enamel Mass.—XIII., Appliances for Smelting the Enamel Mass.—XIV., Smelting the Charge.—XV., Composition of Enamel Masses.—XVI., Composition of Masses for Ground Enamels.—XVII., Composition of Cover Enamels.—XVIII., Preparing the Articles for Enamelling.—XIX., Applying the Enamel.—XX., Firing the Ground Enamel.—XXI., Applying and Firing the Cover Enamel or Glaze.—XXII., Repairing Defects in Enamelled Ware.—XXIII., Enamelling Articles of Sheet Metal.—XXIV., Decorating Enamelled Ware.—XXV., Specialities in Enamelling.—XXVI., Dial-plate Enamelling.—XXVII., Enamels for Artistic Purposes: Recipes for Enamels of Various Colours.—Index.

Books on Textile Subjects.

THE TECHNICAL TESTING OF YARNS AND TEXTILE FABRICS, with Reference to Official Specifications. Translated from the German of Dr. J. HERZFELD. With 69 Illustrations. Price 10s. 6d. ; France and Belgium, 13 frs. ; Colonies, 12s., post free.

Contents.

Yarn Testing. III., Determining the Yarn Number.—IV., Testing the Length of Yarns.—V., Examination of the External Appearance of Yarn.—VI., Determining the Twist of Yarn and Twist.—VII., Determination of Tensile Strength and Elasticity.—VIII., Estimating the Percentage of Fat in Yarn.—IX., **Determination of Moisture** (Conditioning).—Appendix.

Press Opinions.

"The author has endeavoured to collect and arrange in systematic form for the first time all the data relating to both physical and chemical tests as used throughout the whole of the textile industry, so that not only the commercial and textile chemist who has frequently to reply to questions on these matters, but also the practical manufacturer of textiles and his subordinates, whether in spinning, weaving, dyeing, and finishing, are catered for. . . . The book is profusely illustrated, and the subjects of these illustrations are clearly described."—*Textile Manufacturer.*

"This is probably the most exhaustive book published in English on the subject dealt with. . . . We have great confidence in recommending the purchase of this book by all manufacturers of textile goods of whatever kind, and are convinced that the concise and direct way in which it is written, which has been admirably conserved by the translator, renders it peculiarly adapted for the use of English readers."—*Textile Recorder.*

"A careful study of this book enables one to say with certainty that it is a standard work on the subject. Its importance is enhanced greatly by the probability that we have here, for the first time in our own language, in one volume, a full, accurate, and detailed account, by a practical expert, of the best technical methods for the testing of textile materials, whether in the raw state or in the more or less finished product."—*Glasgow Herald.*

"It would be well if our English manufacturers would avail themselves of this important addition to the extensive list of German publications which, by the spread of technical information, contribute in no small degree to the success, and sometimes to the supremacy, of Germany in almost every branch of textile manufacture."—*Manchester Courier.*

DECORATIVE AND FANCY TEXTILE FABRICS.

With Designs and Illustrations. By R. T. LORD. A Valuable Book for Manufacturers and Designers of Carpets, Damask, Dress and all Textile Fabrics. Price 7s. 6d.; Other Countries, 9s., post free.

Contents.

Chapters I., A few Hints on Designing Ornamental Textile Fabrics.—II., A few Hints on Designing Ornamental Textile Fabrics (continued).—III., A few Hints on Designing Ornamental Textile Fabrics (continued).—IV., A few Hints on Designing Ornamental Textile Fabrics (continued).—V., Hints for Ruled-paper Draughtsmen.—VI., The Jacquard Machine.—VII., Brussels and Wilton Carpets.—VIII., Tapestry Carpets.—IX., Ingrain Carpets.—X. Axminster Carpets.—XI., Damask and Tapestry Fabrics.—XII., Scarf Silks and Ribbons.—XIII., Silk Handkerchiefs.—XIV., Dress Fabrics.—XV., Mantle Cloths.—XVI., Figured Plush.—XVII., Bed Quilts.—XVIII., Calico Printing.

Press Opinions.

"The book is to be commended as a model manual, appearing at an opportune time, since every day is making known a growing desire for development in British industrial art."—*Dundee Advertiser.*

"Those engaged in the designing of dress, mantle tapestry, carpet and other ornamental textiles will find this volume a useful work of reference."—*Leeds Mercury.*

"The writer's avocation is that of a designer for the trade, and he therefore knows what he is writing about. . . . The work is well printed and abundantly illustrated, and for the author's share of the work we have nothing but commendation. It is a work which the student designer will find thoroughly useful."—*Textile Mercury.*

"Designers especially, who desire to make progress in their calling, will do well to take the hints thrown out in the first four chapters on 'Designing Ornamental Textile Fabrics'."—*Nottingham Daily Guardian.*

"The book can be strongly recommended to students and practical men."—*Textile Colorist.*

POWER-LOOM WEAVING AND YARN NUMBERING,

according to various Systems, with Conversion Tables. An Auxiliary and Text-book for Pupils of Weaving Schools, as well as for self-instruction and for general use, by those engaged in the Weaving Industry. Translated from the German of ANTHON GRUNER. With Coloured Diagrams. Price 7s. 6d.; Abroad, 9s., strictly net, post free.

Contents.

I., **Power-Loom Weaving in General.** Various Systems of Looms.—II., **Mounting and Starting the Power-Loom.** English Looms.—Tappet or Treadle Looms.—Dobbies.—III., **General Remarks on the Numbering, Reeling and Packing of Yarn.—Appendix. Useful Hints.** Calculating Warps.—Weft Calculations.—Calculations of Cost Price in Hanks

Press Opinions.

"This work brings before weavers who are actually engaged in the various branches of fabrics, as well as the technical student, the different parts of the general run of power-looms in such a manner that the parts of the loom and their bearing to each other can be readily understood.... The work should prove of much value, as it is in every sense practical, and is put before the reader in such a clear manner that it can be easily understood."—*Textile Industries.*

"The work has been clearly translated from the German and published with suitable illustrations.... The author has dealt very practically with the subject."—*Bradford Daily Telegraph.*

"The book, which contains a number of useful coloured diagrams, should prove invaluable to the student, and its handy form will enable it to become a companion more than some cumbrous work."—*Cotton Factory Times.*

"The book has been prepared with great care, and is most usefully illustrated. It is a capital text-book for use in the weaving schools or for self-instruction, while all engaged in the weaving industry will find its suggestions helpful."—*Northern Daily Telegraph.*

THE COLOUR PRINTING OF CARPET YARNS.

A Useful Manual for Colour-Chemists and Textile Printers, by DAVID PATERSON, F.C.S. 132 pp. Illustrated. Price 7s. 6d.; Abroad, 9s., strictly net, post free.

Contents.

Chapters I., Structure and Constitution of Wool Fibre.—II., Yarn Scouring.—III., Scouring Materials.—IV., Water for Scouring.—V., Bleaching Carpet Yarns.—VI., Colour Making for Yarn Printing.—VII., Colour Printing Pastes.—VIII., Colour Recipes for Yarn Printing.—IX., Science of Colour Mixing.—X., Matching of Colours.—XI., "Hank" Printing.—XII., Printing Tapestry Carpet Yarns.—XIII., Yarn Printing.—XIV., Steaming Printed Yarns.—XV., Washing of Steamed Yarns.—XVI., Aniline Colours suitable for Yarn Printing.—XVII., Glossary of Dyes and Dye-wares used in Wood Yarn Printing.—Appendix.

Press Opinions.

"The subject is very exhaustively treated in all its branches.... The work, which is very well illustrated with designs, machines, and wool fibres, will be a useful addition to our textile literature."—*Northern Whig.*

"The book is worthy the attention of the trade."—*Worcester Herald.*

"An eminent expert himself, the author has evidently strained every effort in order to make his work the standard guide of its class"—*Leicester Post.*

"It gives an account of its subject which is both valuable and instructive in itself, and likely to be all the more welcome because books dealing with textile fabrics usually have little or nothing to say about this way of decorating them."—*Scotsman.*

"The work shows a thorough grasp of the leading characteristics as well as the minutiæ of the industry, and gives a lucid description of its chief departments.... As a text-book in technical schools where this branch of industrial education is taught the book is valuable, or it may be perused with pleasure as well as profit by any one having an interest in textile industries."—*Dundee Courier.*

"The treatise is arranged with great care, and follows the processes described in a manner at once clear and convincing."—*Glasgow Record.*

Books on Plumbing.

EXTERNAL PLUMBING WORK.

A Treatise on Lead Work for Roofs. By JOHN W. HART, R.P.C. Price 7s. 6d., post free; Other Countries, 8s.

List of Chapters.

Chapters I., Cast Sheet Lead.—II., Milled Sheet Lead.—III., Roof Cesspools.—IV., Socket Pipes.—V., Drips.—VI., Gutters.—VII., Gutters (continued).—VIII., Breaks.—IX., Circular Breaks.—X., Flats.—XI., Flats (continued).—XII., Rolls on Flats.—XIII., Roll Ends.—XIV., Roll Intersections.—XV., Seam Rolls.—XVI., Seam Rolls (continued).—XVII., Tack Fixings.—XVIII., Step Flashings.—XIX., Step Flashings (continued).—XX., Secret Gutters.—XXI., Soakers.—XXII., Hip and Valley Soakers.—XXIII., Dormer Windows.—XXIV., Dormer Windows (continued).—XXV., Dormer Tops.—XXVI., Internal Dormers.—XXVII., Skylights.—XXVIII., Hips and Ridging.—XXIX., Hips and Ridging (continued).—XXX., Fixings for Hips and Ridging.—XXXI., Ornamental Ridging.—XXXII., Ornamental Curb Rolls.—XXXIII., Curb Rolls.—XXXIV., Cornices.—XXXV., Towers and Finials.—XXXVI., Towers and Finials (continued).—XXXVII., Towers and Finials (continued).—XXXVIII., Domes.—XXXIX., Domes (continued).—XL., Ornamental Lead Work.—XLI., Rain Water Heads.—XLII., Rain Water Heads (continued).—XLIII., Rain Water Heads (continued).

Press Opinions.

"The publication of this book will do much to stimulate attention and study to external plumbing work, for it is a book which we can heartily recommend to every plumber, both old and young, who desires to make himself proficient in the several branches of his trade. We can heartily recommend the book to plumbers and architects."—*Sanitary Record.*

"This is an eminently practical and well-illustrated volume on the management of external lead work."—*Birmingham Daily Post.*

"It is thoroughly practical, containing many valuable hints, and cannot fail to be of great benefit to those who have not had large experience."—*Sanitary Journal.*

"With Mr. Hart's treatise in his hands the young plumber need not be afraid of tackling outside work. He would do well to study its pages at leisure, so that he may be ready for it when called upon."—*Ironmongery.*

"Works on sanitary plumbing are by no means rare, but treatises dealing with external plumbing work are sufficiently scarce to ensure for Mr. Hart's new publication a hearty reception."—*The Ironmonger.*

HINTS TO PLUMBERS ON JOINT WIPING, PIPE BENDING AND LEAD BURNING. Second Edition, Revised and Corrected. By JOHN W. HART, R.P.C. Over 300 pages, Illustrated. Price 7s. 6d.; Other Countries, 8s., post free.

List of Chapters.

x., Introduction.—Chapters I., Pipe Bending.—II., Pipe Bending (continued).—III., Pipe Bending (continued).—IV., Square Pipe Bendings.—V., Half-circular Elbows.—VI., Curved Bends on Square Pipe.—VII., Bossed Bends.—VIII., Curved Plinth Bends.—IX., Rain-water Shoes on Square Pipe.—X., Curved and Angle Bends.—XI., Square Pipe Fixings.—XII., Joint-wiping.—XIII., Substitutes for Wiped Joints.—XIV., Preparing Wiped Joints.—XV., Joint Fixings.—XVI., Plumbing Irons.—XVII., Joint Fixings.—XVIII., Use of "Touch" in Soldering.—XIX., Underhand Joints.—XX., Blown and Copper Bit Joints.—XXI., Branch Joints.—XXII., Branch Joints (continued).—XXIII., Block Joints.—XXIV., Block Joints (continued).—XXV., Block Fixings.—XXVI., Astragal Joints—Pipe Fixings.—XXVII., Large Branch Joints.—XXVIII., Large Underhand Joints.—XXIX., Solders.—XXX., Autogenous Soldering or Lead Burning.

Press Opinions.

"Rich in useful diagrams as well as in hints."—*Liverpool Mercury.*

"A well got-up and well-done practical book. It is freely illustrated and is a reliable help in respect of some of the most awkward work the young plumber has to perform."—*The Ironmonger.*

"The papers are eminently practical, and go much further into the mysteries they describe than the title 'Hints' properly suggests."—*Scotsman.*

"The articles are apparently written by a thoroughly practical man. As a practical guide the book will doubtless be of much service."—*Glasgow Herald.*

"So far as the practical hints in this work are concerned, it will be useful to apprentices and students in technical schools, as it deals mainly with the most important or difficult branches of the plumber's craft, *viz.*, joint wiping, pipe bending and lead burning. . . . 'Hints' are the most useful things to an apprentice, and there are many in this work which are not to be found in some of the text-books."—*English Mechanic.*

"It is a book for the intelligent operative first of all, not a mere manual of instruction for the beginner, nor yet a scientific treatise on the whole art of sanitary plumbing. The special subject with which it deals is joint-making, the most important branch of the operative's work, and into this topic the author goes with a thoroughness that is full of suggestion to even the most experienced workman. There is no one who has to do with plumbing but could read the book with profit."—*Ironmongery.*

WORKS IN PREPARATION

A HISTORY OF DECORATIVE ART. For Designers, Decorators and Workmen. [*Nearly Ready*.

HOUSE PAINTING AND DECORATING. A Handbook for Painters and Decorators. [*Nearly Ready*.

THE PRINCIPLES AND PRACTICE OF DIPPING, BURNISHING AND BRONZING BRASS WORK. [*Nearly Ready*.

THE SCIENCE OF COLOUR MIXING. A Manual intended for the use of Dyers, Calico Printers, Colour Chemists and Students. By DAVID PATERSON, F.C.S.

WAXES.

AGRICULTURAL CHEMISTRY.

THE MANUFACTURE OF BRUSHES OF EVERY DESCRIPTION.

THE ART AND PRACTICE OF BLEACHING.
[*In the Press*.

THE MANUFACTURE OF LEATHER. Translated from the French of M. VILLON. [*In the Press*.

A TREATISE ON THE CERAMIC INDUSTRY. By EMILLE BOURRY.

MINING SAFETY APPLIANCES.

THE MANUFACTURE OF LAKE PIGMENTS. By T. H. JENNISON, F.S.S., etc. [*In the Press*.

THE RISKS AND DANGERS OF VARIOUS OCCUPATIONS AND THEIR PREVENTIONS. By Dr. L. A. PARRY.

COLOUR MATCHING ON TEXTILES. A Manual intended for the use of Students of Colour Chemistry, Dyeing and Textile Printing. By DAVID PATERSON, F.C.S.

TERRA-COTTA, BRICKS AND POTTERY FOR BUILDING PURPOSES.

TECHNOLOGY OF PETROLEUM. By NEUBURGER and NOALHAT.

HOT WATER SUPPLY. [*In the Press.*

THE CULTURE OF HOPS. [*In the Press.*

THE RONTJEN RAYS IN MEDICAL PRACTICE.

CONTINENTAL PATENTS FOR GAS APPARATUS.

SULPHATES OF IRON AND ALUMINIUM AND ALUM INDUSTRY. By L. GESCHWIND.

SCOTT, GREENWOOD & CO.

are Publishers of the following old-established and well-known Trade Journals:—

THE OIL AND COLOURMAN'S JOURNAL. The Organ of the Oil, Paint, Drysaltery and Chemical Trades. Home Subscription, 7s. 6d. per year; United States, $2; Other Countries, 10s. per year.

THE POTTERY GAZETTE. For the China and Glass Trades. Home Subscription, 7s. 6d. per year; United States, $2; Other Countries, 10s. per year.

THE HATTERS' GAZETTE. Home Subscription, 6s. 6d. per year; United States, $2; Other Countries, 9s. per year.

THE DECORATORS' GAZETTE AND PLUMBERS' REVIEW. Home Subscription, 6s. 6d. per year; United States, $2; Other Countries, 9s.

19 Ludgate Hill, London, E.C.

24/4/1900.

www.ingramcontent.com/pod-product-compliance
Lightning Source LLC
Chambersburg PA
CBHW021809230426
43669CB00008B/686